The Roadmap to Renewable Energy

Disha Gupta • Daniel J. Soeder • Aditya Chichani

The Roadmap to Renewable Energy

Integrating Science, Business, and AI for a Sustainable Future

Springer

Disha Gupta
Geology and Geological Engineering
EDP Renewables North America
Houston, TX, USA

Daniel J. Soeder
Geology and Geological Engineering
Soeder Geoscience LLC
Tunnelton, WV, USA

Aditya Chichani
Computer Science
Walmart Labs
Milpitas, CA, USA

ISBN 978-3-032-09608-1 ISBN 978-3-032-09609-8 (eBook)
https://doi.org/10.1007/978-3-032-09609-8

This Springer imprint is published by the registered company Springer Nature Switzerland AG
The registered company address is: Gewerbestrasse 11, 6330 Cham, Switzerland

Preface

The global transition to clean energy is not merely a technological shift, it is a transformation that requires the seamless integration of science, strategic business planning, and intelligent data-driven decision-making. As the world confronts the dual imperatives of climate change and energy equity, the role of renewable energy has never been more critical.

Roadmap to Renewable Energy: Integrating Science, Business, and AI for a Sustainable Future is a comprehensive guide born out of the real-world challenges and solutions encountered in the development of renewable energy projects. Drawing from my experience leading large-scale wind, solar, and battery storage initiatives across the United States, this book demystifies the process of bringing a renewable energy project from inception to reality.

What sets this book apart is its interdisciplinary approach. It bridges the gap between geoscience and project finance, between regulatory insight and technological innovation, and between strategic planning and artificial intelligence. With contributions from Dan Soeder, a respected geoscientist and former US Department of Energy researcher, and Aditya Chichani, an accomplished AI/ML expert specializing in large-scale search and recommendation systems, we bring diverse yet harmonized perspectives to the energy conversation.

This book is designed for students, professionals, developers, policy leaders, and anyone seeking a deeper understanding of how renewable energy projects are conceptualized, financed, executed, and optimized. By uniting scientific rigor with business pragmatism and the transformative power of AI, we aim to equip readers with both the knowledge and the vision to accelerate the global clean energy transition.

Houston, TX, USA
May 2025

Disha Gupta

Acknowledgements This book is the product of countless conversations, collaborations, and quiet moments of reflection, and I owe its completion to many who stood by me throughout this journey.

My sincere thanks go to my co-authors. Dan, your decades of experience in earth science, energy systems, and environmental science have shaped not only the scientific foundation of this book but also my own understanding of the renewable energy landscape. Your generous mentorship, intellectual clarity, and ability to translate complex concepts into practical insight elevated this work beyond what I imagined possible. Collaborating with you has been a learning experience I will always value.

Aditya, thank you not only for your brilliant contributions to this book but also for your thoughtfulness, patience, and perspective. Your role in refining the intersection of artificial intelligence and renewable energy in this book added a critical dimension to our message.

I also wish to extend my deepest gratitude to my mom and dad, whose unwavering love, resilience, and encouragement have been my foundation. To my brother, thank you for being a steady source of support and belief, even during the most challenging stretches of this endeavor.

Lastly, to all the mentors, colleagues, and readers who have inspired this work, your presence has made this journey richer and more purposeful than I could have imagined.

Disha Gupta

Contents

Chapter 1
Introduction

1.1 Overview of Renewable Energy

Global energy resources have been predominantly based on fossil fuels since the Industrial Revolution, particularly following the end of World War II [1]. Prior to this, societies were constrained to localized energy forms: human and animal labor, wood combustion for heating and cooking, and limited mechanical energy from water wheels and windmills. Although ancient civilizations were aware of fossil fuels such as coal, natural gas, and petroleum, these resources were not extensively exploited for energy purposes until much later [1].

The accelerated adoption of fossil fuels dramatically reshaped human civilization. Coal, the first fossil fuel widely used for industrial purposes, powered steam engines, manufacturing processes, and transportation networks [1]. Petroleum subsequently emerged as a major energy source following the refinement of kerosene and the advent of the internal combustion engine, which revolutionized personal and commercial transportation. Natural gas, often found alongside coal and oil deposits, added to the energy portfolio by providing a cleaner-burning alternative for heating and electricity generation [1].

Fossil fuels provided a concentrated, portable, and relatively inexpensive form of energy that enabled rapid industrialization, urbanization, and technological advancement. However, their extraction and combustion also introduced profound environmental consequences. Chief among these is the release of carbon dioxide (CO_2), a greenhouse gas that traps heat in the atmosphere and contributes to global warming [1]. Other impacts include ecosystem disruption from resource extraction, ocean acidification resulting from CO_2 absorption, and severe environmental degradation from oil spills and coal mining operations.

The finite nature of fossil fuel reserves adds another layer of urgency. Coal, petroleum, and natural gas are non-renewable on human timescales; once depleted, they cannot be replaced. Moreover, continued reliance on fossil fuels exacerbates

D. Gupta et al., *The Roadmap to Renewable Energy*,
https://doi.org/10.1007/978-3-032-09609-8_1

geopolitical tensions, as energy security becomes intertwined with global supply chains and international politics [1].

Renewable energy sources such as wind, solar, geothermal, hydroelectric, and biofuels offer a sustainable alternative. Unlike fossil fuels, renewables are naturally replenished and generate minimal greenhouse gas emissions during use [1]. Advances in technology have made renewable energy more economically competitive, and ongoing research continues to improve efficiency, storage capabilities, and grid integration.

The shift towards integrating multiple renewable technologies into a cohesive and resilient energy framework is essential for addressing climate change, ensuring long-term energy security, and promoting environmental sustainability. This transition requires not only technological innovation but also systemic changes in infrastructure, policy, and public behavior [1].

1.1.1 Global Energy and the History of Fossil Fuels

Before the Industrial Revolution, energy came mainly from human and animal labor, wood or charcoal combustion, and mechanical power from water wheels and windmills [1]. Although fossil fuels were known, they were rarely used.

The first significant use of coal occurred in seventeenth-century England when a wood shortage prompted industries, particularly glassmakers, to search for alternatives [1]. King James I reserved wood for shipbuilding and prohibited its use for melting glass, compelling glassmakers to switch to coal. They found that coal not only burned hotter than wood but also allowed the production of stronger and thicker glass. This innovation contributed to the popularity of British glass across Europe, particularly in France, where robust bottles were needed to contain the newly invented champagne [1].

The advantages of coal's high combustion temperature and cost-effectiveness soon attracted blacksmiths, masons, and other industries. As the Industrial Revolution gained momentum in the mid-eighteenth century, coal became essential for powering steam engines, factories, ships, and railroads. By the early twentieth century, most machinery had transitioned to electricity, but coal continued to dominate electricity generation [1].

Coal is formed from the preserved remains of woody plants that lived millions of years ago in swampy environments. These plants were buried under sediment in oxygen-poor (anoxic) conditions, preventing decay. Over geological time, the organic material underwent thermal maturation under heat and pressure, evolving through stages from peat to lignite, bituminous coal, and finally anthracite, the purest form of coal [1]. The energy stored in coal originates from ancient photosynthesis, where plants captured solar energy and fixed carbon dioxide into organic structures.

Today, China is the world's largest coal producer, mining and consuming more coal annually than the next nine largest producers combined. India follows in second place, driven by its rapid industrialization and electrification, while the United States, although once a global leader, has seen declining coal production since 2006 due to competition from abundant domestic shale gas. Australia ranks fourth in production but exports most of its coal, especially to China. Other significant producers include Indonesia, Russia, South Africa, Germany, and Poland [1].

While coal's dominance was established early, petroleum emerged centuries later as a major energy resource. Known to ancient civilizations primarily for medicinal purposes, petroleum was long considered a "rock oil" and used sparingly for treating illnesses [1]. The shift toward petroleum as an energy source began in the mid-nineteenth century when Abraham Gesner developed a method for refining petroleum into kerosene, a clean-burning lamp fuel. Samuel Kier commercialized kerosene in the United States, designing lamps specifically to burn it and thus helping to drive whale oil toward obsolescence [1].

Kier's kerosene lanterns, coupled with Edwin Drake's successful oil drilling operation in Pennsylvania in 1859, launched the modern petroleum industry. The Spindletop gusher in Texas in 1901 dramatically demonstrated the vast potential of oil extraction, producing more petroleum from a single well than previously seen from entire fields [1]. Petroleum quickly found a growing market with the rise of the automobile, especially after Henry Ford's Model T made personal vehicles accessible to the masses.

Petroleum, like coal, originates from ancient organic material but derives primarily from microscopic algae and soft plants deposited in anoxic marine environments. Over millions of years, thermal maturation converted this organic matter into hydrocarbons. Different types of kerogen, which is the insoluble organic matter in sedimentary rocks, produce different fossil fuels depending on their biological source material and the temperature and pressure conditions they experienced [1]. Type I and II kerogens, formed from freshwater and marine algae respectively, are oil-prone, while Type III kerogen, formed from woody land plants, is coal-prone.

Natural gas, consisting mainly of methane, is commonly found associated with oil or coal deposits. It forms either biogenically, through microbial activity at shallow depths, or thermogenically, through deeper thermal maturation processes. Gas resources have increasingly gained attention as a relatively cleaner bridge fuel toward decarbonization.

Thus, the historical trajectory of energy has been defined by the discovery, extraction, and consumption of fossil fuels, starting with coal and followed by petroleum and natural gas. These resources powered industrial and societal transformations but also laid the groundwork for the environmental and climate challenges faced today.

1.1.2 The Energy Crisis and Sustainable Solutions

The economic growth driven by fossil fuels also introduced strategic vulnerabilities. These vulnerabilities became dramatically evident during the 1973 Arab oil embargo, where members of the Organization of Petroleum Exporting Countries

(OPEC) restricted oil shipments to nations supporting Israel during the Yom Kippur War [1]. Although the embargo reduced oil supply by only about 10%, it triggered widespread panic, quadrupling oil prices and causing severe economic disruptions. Gasoline shortages, rationing, and long lines at fuel stations became common across the United States.

Recognizing the national security risks posed by heavy dependence on foreign petroleum, the U.S. Congress created the Department of Energy (DOE) in 1977. The DOE consolidated numerous federal energy-related functions, aiming to coordinate national energy policy, encourage diversification of energy sources, and promote energy conservation and technological innovation. One major goal was to reduce U.S. reliance on foreign oil by developing domestic energy resources, including coal, nuclear power, and alternative fuels [1].

Technological innovations subsequently altered the energy landscape. Techniques such as horizontal drilling and hydraulic fracturing, pioneered by George Mitchell and others, enabled access to vast reserves of oil and natural gas trapped in low-permeability shale formations [1]. These breakthroughs in the 1990s and 2000s led to a boom in domestic energy production, allowing the United States to significantly reduce its dependence on imported petroleum.

While increased domestic fossil fuel production temporarily bolstered energy security, it did not eliminate the environmental and climate impacts associated with fossil fuel use. The combustion of coal, oil, and natural gas remains the primary source of anthropogenic carbon dioxide (CO_2) emissions, contributing to global warming and climate change [1]. Moreover, the extraction, processing, and transportation of fossil fuels present additional environmental hazards. Notable incidents include oil spills, such as the 2010 Deepwater Horizon disaster in the Gulf of Mexico, which caused widespread ecological damage [2].

Another significant yet less visible consequence of rising atmospheric CO_2 levels is ocean acidification. When CO_2 dissolves in seawater, it forms carbonic acid, lowering ocean pH and threatening marine organisms that depend on calcium carbonate for their shells and skeletons [1]. This process jeopardizes entire marine food webs and the human economies that rely upon them.

Despite the clear environmental risks, fossil fuels continue to dominate global energy consumption because of existing infrastructure investments, economic inertia, and the high energy density of fossil resources [1]. However, the mounting evidence of climate change, ecological degradation, and resource depletion underscores the urgent need to transition toward more sustainable energy systems.

1.1.3 Types of Renewable Energy

Renewable energy encompasses a range of naturally replenished resources that offer sustainable alternatives to fossil fuels. These technologies harness the energy of the sun, wind, biological processes, the Earth's heat, and the oceans. Although renewables present distinct technical challenges, their role is increasingly critical in mitigating climate change, enhancing energy security, and supporting long-term economic stability [1].

1.1.3.1 Wind Energy

Wind energy has become one of the fastest-growing renewable energy sources globally. Modern wind turbines, typically configured with three blades and horizontal-axis designs, have a rated capacity of between two to four megawatts (MW) each [1]. Offshore wind farms, particularly in Europe, exploit stronger and more consistent marine winds, though installation and maintenance costs are higher compared to land-based systems. Wind turbines require large surface areas for optimal operation but produce no direct greenhouse gas emissions during use. However, integrating wind power into existing electrical grids necessitates careful management due to variability and intermittency.

1.1.3.2 Solar Energy

Solar power technologies capture sunlight either through photovoltaic (PV) cells or through solar thermal systems. PV cells, made from semiconductor materials, directly convert sunlight into electricity via the photoelectric effect [1]. Solar thermal plants, in contrast, concentrate sunlight using mirrors or lenses to produce heat, which can then be used to generate electricity through conventional steam turbines. The declining costs of solar panels, coupled with improvements in efficiency, have accelerated global adoption. Nevertheless, solar generation is inherently intermittent and depends on weather conditions and daylight availability, necessitating advances in energy storage and grid flexibility.

1.1.3.3 Biofuels

Biofuels represent renewable energy derived from biological materials. These include methane (biogas) generated from organic waste, ethanol produced by fermenting corn or sugarcane, and biodiesel derived from vegetable oils or animal fats [1]. Moreover, when evaluating the sustainability of biofuels, it is crucial to consider their full lifecycle greenhouse gas (GHG) emissions, from cultivation, harvesting, processing, and transportation to combustion. Some biofuels may reduce emissions compared to fossil fuels, but others may offer limited or even negative climate benefits if associated land-use changes and fossil fuel inputs are high. Therefore, careful management and a shift toward waste-based or non-food feedstocks are essential to maximizing the climate advantages of biofuels.

1.1.3.4 Hydrogen

Hydrogen is an energy carrier with significant potential to decarbonize various sectors. When produced from renewable electricity through water electrolysis (a process splitting water into hydrogen and oxygen), it is termed "green hydrogen" [1]. Hydrogen can be combusted directly or converted into electricity using fuel cells, emitting only water vapor as a byproduct. Its high energy density makes hydrogen particularly attractive for applications where batteries are less practical, such as heavy transport and industrial processes. However, hydrogen's current production is dominated by fossil-fuel-based methods, and building the necessary renewable hydrogen infrastructure remains a major challenge.

1.1.3.5 Ocean Energy

Ocean energy technologies tap into the kinetic and thermal energy of the oceans. Approaches include tidal power, wave energy converters, and ocean thermal energy conversion systems [1]. Japan and other island nations have invested significantly in ocean energy research. Despite immense theoretical potential, ocean energy remains in the early stages of commercialization due to high costs, technical difficulties, and environmental concerns related to marine ecosystems.

1.1.3.6 Geothermal Energy

Geothermal energy exploits the Earth's internal heat, accessible through hot water reservoirs, steam fields, or enhanced geothermal systems where water is injected into hot rock formations [1]. Geothermal plants produce baseload electricity with minimal surface footprint and very low greenhouse gas emissions. Countries like Iceland, New Zealand, and Kenya have successfully integrated geothermal into their national grids. Technological innovations such as improved drilling techniques could significantly expand geothermal capacity, even in regions without naturally favorable conditions.

Each renewable energy source presents unique opportunities and challenges, but collectively, they offer a diverse and resilient pathway to reduce dependence on finite fossil resources and achieve deep decarbonization.

1.1.4 Applications to Electricity

Electricity plays a critical role in modern society, yet it is not a primary energy source; rather, it is a secondary form of energy derived from the conversion of primary sources such as coal, natural gas, nuclear, wind, and solar power [1]. Electricity offers a flexible and efficient means of transporting and utilizing energy for residential, commercial, and industrial needs. However, as of the early 2020s, the majority of global electricity generation continues to rely heavily on fossil fuels, with coal and natural gas remaining dominant sources [1].

Thermoelectric power plants are the primary mechanism for fossil-based electricity generation. These facilities convert the chemical energy of fossil fuels into thermal energy through combustion, which then heats water to create steam. The steam drives turbines connected to electrical generators, completing the conversion to electricity [1]. The most common thermodynamic cycles employed in these plants are the Rankine cycle, used in coal, natural gas steam plants, and nuclear reactors, and the Brayton cycle, which powers simple natural gas combustion turbines.

A more efficient configuration known as the Combined Cycle (CC) power plant integrates both the Brayton and Rankine cycles. In this design, natural gas is combusted in a turbine (Brayton cycle) to generate electricity, and the hot exhaust gases are then used to produce steam that drives a secondary steam turbine (Rankine cycle) [1]. Combined Cycle plants can achieve efficiencies exceeding 60%, substantially higher than conventional single-cycle plants.

Despite improvements in efficiency, Combined Cycle plants remain fundamentally reliant on fossil fuels and contribute to greenhouse gas emissions. Transitioning to renewable electricity requires significant infrastructure changes, given that renewable energy sources such as wind and solar are geographically distributed, intermittent, and lower in energy density compared to fossil fuels [1].

One promising transitional strategy involves retrofitting existing thermoelectric plants. Rather than abandoning or demolishing these facilities, they can be modified to use alternative heat sources like biogas or geothermal energy in place of fossil fuels [1]. For example, if biogas derived from organic waste is combusted instead of natural gas, or if geothermal reservoirs are tapped for steam production, the basic turbine-generator systems of conventional plants can be preserved. This approach leverages existing infrastructure, reduces costs, and accelerates the decarbonization of electricity generation without waiting for the complete buildout of renewable-specific generation and storage technologies.

As renewable energy sources continue to expand, maintaining grid reliability will require a mix of flexible generation, storage technologies, smart grid innovations, and policy frameworks that encourage investment in sustainable electricity systems.

1.1.5 Applications to Vehicles

Transportation represents one of the most challenging sectors to decarbonize, accounting for a significant share of global greenhouse gas emissions. Various renewable energy approaches offer potential pathways to reduce the carbon footprint of vehicles, including electric vehicles (EVs), biofuel-powered internal combustion engines, biogas vehicles, and hydrogen fuel cell electric vehicles (FCEVs) [1].

1.1.5.1 Electric Vehicles (EVs)

Electric vehicles store electrical energy in rechargeable batteries, typically lithium-ion, and use it to power electric motors for propulsion. EVs offer zero tailpipe emissions, substantially reducing urban air pollution compared to traditional gasoline or diesel-powered vehicles [1]. However, the environmental benefits of EVs depend heavily on the carbon intensity of the electricity used for charging. If the electrical grid is powered predominantly by fossil fuels, the lifecycle emissions of EVs are reduced but not eliminated. Additionally, EV adoption faces challenges related to battery cost, energy density, raw material sourcing (e.g., lithium, cobalt), and charging infrastructure availability.

Advancements in battery technologies, including the development of solid-state batteries, promise to improve energy density, reduce charging times, and extend vehicle range, further enhancing the appeal of EVs as a sustainable transportation solution [1].

1.1.5.2 Biofuels

Biofuels such as ethanol and biodiesel offer a renewable alternative to petroleum-based fuels. Ethanol, commonly produced from corn or sugarcane, can be blended with gasoline or used directly in flex-fuel vehicles [1]. Biodiesel, derived from vegetable oils or animal fats, can substitute for petroleum diesel in compression ignition engines. These biofuels participate in a near-closed carbon cycle, as the CO_2 released during combustion was recently absorbed from the atmosphere during feedstock growth.

However, large-scale biofuel production can introduce unintended environmental consequences, such as deforestation, habitat loss, and competition with food crops for arable land [1]. Careful management and the use of waste-based feedstocks are necessary to maximize the climate benefits of biofuels.

1.1.5.3 Biogas Vehicles

Biogas, primarily composed of methane, is produced through the anaerobic digestion of organic matter such as agricultural waste, sewage, and landfill material. Vehicles originally designed for gasoline operation can be modified to run on compressed biogas or natural gas, offering a relatively low-cost and rapid pathway to reducing transportation emissions [1]. Bi-fuel systems, allowing vehicles to switch between gasoline and biogas, provide operational flexibility. Biogas derived from waste sources is considered carbon-neutral, contributing to both emissions reductions and waste management.

1.1.5.4 Hydrogen Fuel Cell Vehicles (FCEVs)

Hydrogen fuel cell electric vehicles generate electricity onboard by combining hydrogen gas with oxygen from the air in an electrochemical reaction, producing only water vapor as a byproduct [1]. FCEVs offer longer driving ranges and faster refueling times compared to battery-electric vehicles, making them attractive for heavy-duty transport and long-distance travel.

Nevertheless, the widespread adoption of FCEVs faces several obstacles. Most commercial hydrogen is currently produced from natural gas through steam methane reforming, which emits CO_2. Transitioning to green hydrogen produced via renewable-powered electrolysis remains technologically and economically challenging. Moreover, hydrogen infrastructure, including production, transport, and refueling stations, is still underdeveloped in most regions [1].

Each renewable transportation technology presents trade-offs among cost, convenience, emissions reduction potential, and infrastructure requirements. A diversified approach, leveraging the strengths of different technologies, is likely necessary to achieve a comprehensive transition toward sustainable mobility.

To conclude, the historical reliance on fossil fuels has undeniably powered human advancement, supporting industrialization, urbanization, and improvements in quality of life. However, the environmental, economic, and strategic vulnerabilities of fossil fuel dependence have become increasingly evident. Finite fossil fuel resources, climate change driven by rising carbon dioxide emissions, ecosystem degradation, and geopolitical tensions linked to energy security underscore the urgent need for a global transition to renewable energy [1].

Renewable energy technologies, including wind, solar, biofuels, hydrogen, geothermal, and ocean energy, offer viable pathways to a sustainable energy future. These resources are naturally replenished on human timescales and generate minimal greenhouse gas emissions during operation. Although challenges related to intermittency, energy density, infrastructure, and cost persist, ongoing technological innovations continue to improve the performance, affordability, and scalability of renewable systems [1].

Addressing the climate crisis and achieving a sustainable energy future will require more than technological innovation alone. Systemic changes in infrastructure, economic structures, public policy, and societal behavior are essential. Strategic investments in renewable energy generation, storage technologies, smart grids, and supportive regulatory frameworks will be critical to overcoming existing barriers and accelerating the energy transition [1].

Ultimately, societies that embrace the diversification and integration of renewable energy technologies will lead the twenty-first-century energy economy. Those that resist adaptation risk not only environmental and economic costs but also geopolitical marginalization. Renewable energy handbook—the strategic combination of multiple renewable resources, presents a resilient and flexible path forward for building a cleaner, more sustainable, and more secure global energy system.

1.2 An Interdisciplinary Approach to Renewable Energy

The transition to renewable energy is more than a technical or scientific revolution; it is an interdisciplinary endeavor that requires the seamless integration of science, business, and AI. This section explores the critical role of each domain in driving the growth of renewable energy projects, emphasizing how these elements interact and complement one another to create a sustainable and promising energy future.

1.2.1 Integration of Science and Technology

1.2.1.1 The Role of Scientific Research in Advancing Renewable Energy Technologies

The foundation of the renewable energy industry rests on scientific research, which drives the development of advanced technologies and new methodologies. Scientific research is essential in the continuous advancement of renewable energy technologies. Significant innovations in material science have enhanced the performance and reduced the cost of mainstream renewable energy sources. Scientific breakthroughs in photovoltaic cells, wind turbine aerodynamics, and energy storage systems have pushed the boundaries of what is possible. For instance, advancements in materials science have allowed the creation of more efficient and durable solar panels.

These developments in **solar photovoltaic (PV)** technology have led to the emergence of **third-generation solar cells** such as **perovskite solar cells**, **quantum dot cells**, and **organic PVs**. **Perovskite solar cells** have demonstrated promising advancements in efficiency, with power conversion rates increasing from 10% to over 20% in just three years [3]. However, further research is needed to address challenges related to their long-term stability and commercial scalability.

Innovations in aerodynamics and mechanical engineering have optimized wind turbines to harness energy from even low wind speeds [4]. Other renewable technologies, such as **floating offshore wind turbines**, are expanding the potential of wind energy in deeper waters, unlocking new areas for energy production. Research into **biomass energy conversion** and **geothermal energy** is also advancing, contributing to cleaner and more efficient methods of harnessing renewable resources [5]. These examples illustrate how scientific innovation continues to drive the development of renewable energy technologies and create more sustainable solutions.

1.2.1.2 Importance of Technological Innovation in Overcoming Industry Challenges

Technological innovations are crucial in addressing key challenges such as intermittency, storage limitations, and grid integration in renewable energy. Energy storage technologies like lithium-ion and solid-state batteries are bridging the gap between energy generation and demand, solving the challenge of intermittency for solar and wind energy [6]. Other innovations such as **tandem cells**, **multi-junction cells**, and **perovskite-silicon hybrid cells** are enabling higher energy conversion efficiencies while reducing costs. **Concentrated Solar Power (CSP)** systems, which focus sunlight onto a receiver to generate electricity, are also benefiting from innovations in molten salt storage and heat transfer fluids, allowing for continuous power generation even after the sun sets [7].

In wind energy, **floating wind farms** are an emerging solution that allows turbines to be deployed in deep waters, unlocking untapped wind resources. These technologies, along with **vertical axis wind turbines (VAWTs)**, are pushing the boundaries of what wind power can achieve, making it more flexible and adaptable to various locations [5].

Energy storage technologies, such as **lithium-ion batteries**, are another area where innovation is overcoming traditional challenges. Continued improvements in battery lifespan, energy density, and safety are making renewable energy more reliable and cost-effective for large-scale applications. Moreover, **flow batteries** and **hydrogen storage systems** are being researched as alternatives for long-duration storage, further enhancing the potential for renewable energy integration into the grid [7].

In conclusion, cutting-edge research in fields like artificial intelligence and quantum computing promises to further improve grid management, optimize energy storage, and develop next-generation technologies that will accelerate the transition to renewable sources [8].

1.2.2 Business and Market Dynamics

1.2.2.1 How Business Strategies and Market Trends Impact Renewable Energy Project Development

While scientific and technological advancements form the backbone of the renewable energy sector, business strategies and market forces determine the scale and speed of deployment. The successful development of renewable energy projects is not solely dependent on engineering prowess; it hinges on understanding market dynamics, policy shifts, and investment trends. The financial landscape of renewable energy is shaped by regulatory policies like **tax credits**, **subsidies**, and **renewable portfolio standards (RPS)**, which incentivize both producers and consumers to adopt cleaner energy sources [9].

In addition, market dynamics influence project development. The declining cost of renewable energy technologies, particularly solar and wind, has made them increasingly competitive with traditional fossil fuels [10]. Business strategies, such as **power purchase agreements (PPAs)**[1]and **green bonds**,[2] provide renewable energy developers with financial instruments to secure funding and manage risk [11]. Strategic partnerships between private companies, governments, and investors also play a vital role in scaling renewable energy projects, ensuring that they are both economically viable and environmentally sustainable.

1.2.2.2 The Role of Policy, Economics, and Investment in Scaling Renewable Energy

Government policies and incentives play a pivotal role in fostering the growth of renewable energy projects. These policies create stable markets that attract private investment, encouraging the scaling of technologies like solar and wind energy. Furthermore, the declining costs of renewable energy technologies have made them increasingly competitive with traditional fossil fuel-based energy sources, further incentivizing investment. Public-private partnerships and innovative financial instruments, such as green bonds, are also driving growth in the sector [11].

Investments in renewable energy infrastructure are bolstered by global trends toward decarbonization and corporate sustainability goals. The rise of environmental, social, and governance (ESG) criteria in corporate decision-making has led to increased investment in renewable energy projects. Companies that prioritize sustainability are not only reducing their carbon footprints but also gaining competitive advantages in a market increasingly driven by green energy demands [12].

1.2.3 AI and Decision Making

The third pillar in the interdisciplinary approach to renewable energy development is AI, which has emerged as a critical tool for optimizing energy production, improving operational efficiency, and managing risks. In today's data-driven world, decision-making is increasingly informed by sophisticated models and simulations that leverage vast amounts of data.

1.2.3.1 Use of AI in Optimizing Energy Production, Forecasting, and Efficiency

In renewable energy, AI is essential for optimizing the efficiency of energy production systems by forecasting weather patterns, predicting energy output, and

[1] It is a contract between an energy producer and a buyer (often a utility or large corporation) for the long-term purchase of electricity at a fixed price.

[2] These are debt securities issued to raise capital for financing environmentally sustainable projects, such as renewable energy, energy efficiency, and climate adaptation.

managing grid integration. Wind farm operators, for example, use AI to predict wind patterns and adjust turbine operations to maximize energy capture [13]. Similarly, solar power facilities employ AI models that integrate meteorological data to predict solar irradiance and optimize energy generation [14]. This predictive capability ensures that renewable energy projects operate at peak efficiency and can supply the grid with consistent energy, even under fluctuating environmental conditions.

1.2.3.2 Applications of Data Science in Risk Management, Resource Assessment, and Operational Optimization

Risk management and resource assessment are other crucial areas where data science has a profound impact. Renewable energy developers use geospatial data and machine learning algorithms to assess potential sites for solar or wind projects, considering factors such as topography, climate, and proximity to transmission infrastructure [15]. Predictive analytics further help identify risks related to **regulatory changes**, **market fluctuations**, and **technological obsolescence**, enabling project managers to make informed decisions and mitigate risks [16].

Operational optimization is greatly enhanced through the **real-time monitoring** of energy assets. For instance, **digital twin technology**, which creates virtual models of physical assets, allows operators to simulate and predict how different factors affect the performance of renewable energy systems. Predictive maintenance, powered by AI, reduces downtime and operational costs by identifying potential equipment failures before they occur, leading to higher energy yields and more reliable project performance [17].

We will explore the application of AI in greater detail in upcoming chapters, where we delve into advanced predictive models and their crucial roles in the operational management of renewable energy projects.

1.3 Book Structure and Objective

1.3.1 Purpose of the Book

The motivation behind creating this comprehensive guide stems from the rapidly evolving landscape of renewable energy, where traditional silos of knowledge are no longer sufficient to address the complex challenges and opportunities that arise. This book aims to bridge the gap between renewable energy science, business strategies, and AI, providing a unified resource that empowers stakeholders to make informed decisions throughout the project development lifecycle. By highlighting the necessity of an interdisciplinary approach, we recognize that successful renewable energy initiatives require collaboration across various fields, including engineering, finance, policy, and data science. This holistic perspective will foster innovative solutions that not only enhance project viability but also contribute to global sustainability goals.

1.3.2 Overview of Chapters

This book is structured into twelve chapters, each focusing on critical aspects of renewable energy project development:

1. **Introduction**: Establishes the foundational need for integration across disciplines in renewable energy.
2. **The Science of Renewable Energy**: Explores the fundamental technologies, mechanics, and scientific principles behind solar, wind, storage, and geothermal energy systems.
3. **Policy and Regulatory Frameworks for Renewable Energy**: Analyzes the regulatory landscape that shapes renewable energy markets, including successful case studies.
4. **Environment and Social Considerations**:Evaluates the ecological and societal impacts of renewable energy projects, emphasizing assessments, regulations, and community engagement.
5. **Technological Innovations in Renewable Energy**: Highlights recent advancements in storage, smart grids, and integration technologies along with wind and solar resource assessment that enhance project efficiency.
6. **Financial Modeling and Economic Analysis**: Covers key financial metrics, risk analysis, funding mechanisms, and revenue models essential for evaluating and financing renewable energy projects.
7. **Market Analysis and Business Strategies**: Examines global and regional renewable energy markets, business models, competitive dynamics, and strategic growth opportunities.
8. **Data Science and Machine Learning in Renewable Energy**: Provides a hands-on roadmap for applying the full machine learning lifecycle to solve real-world renewable energy challenges.
9. **Project Management and Implementation**: Outlines the full project lifecycle, from feasibility and design to construction, budgeting, and quality control in renewable energy developments..
10. **Future Trends and Opportunities in Renewable Energy**: Highlights the future of renewable energy, including breakthrough technologies, industrial decarbonization, AI, and opportunities in emerging markets.
11. **Case Studies of Successful Hybrid Renewable Energy Systems**: Offers in-depth analyses of exemplary projects that successfully integrate science, business, and AI, drawing valuable lessons.
12. **Conclusion**: Summarizes key insights and emphasizes the importance of interdisciplinary collaboration, while discussing future trends and challenges in the sector.

1.3.3 Target Audience and Expected Outcomes

This book is designed for a diverse target audience, including professionals in the renewable energy sector, students pursuing related fields, and policymakers involved in energy regulation and planning. By engaging with this material, readers will gain a comprehensive understanding of the multifaceted nature of renewable energy projects. They will be equipped with practical knowledge and strategic insights that facilitate informed decision-making and foster innovative approaches to project development. Ultimately, readers will emerge with the tools necessary to navigate the complexities of the renewable energy landscape, contributing to sustainable energy solutions in their respective roles.

References

1. D.J. Soeder, S.J. Borglum, *The Fossil Fuel Revolution: Shale Gas and Tight Oil* (Elsevier Publishing, Cambridge, 2019)
2. J.L. Ramseur, Deepwater horizon oil spill: recent activities and ongoing developments. Technical Report 7-5700, Congressional Research Service (2015)
3. Massachusetts Institute of Technology, Perovskite solar cells: Challenges and opportunities. Institutional report (2015)
4. S.E. Hosseini, A.M. Andwari, M.A. Wahid, G. Bagheri, A review on green energy potentials in renewable energy technologies. Energy Procedia **74**, 3–10 (2021)
5. United Nations Conference on Trade and Development (UNCTAD), The role of renewable energy in achieving sustainable development goals (2019)
6. X. Luo, J. Wang, M. Dooner, J. Clarke, Overview of current development in electrical energy storage technologies and the application potential in power systems. Renew. Sustain. Energy Rev. **42**, 393–431 (2015)
7. ENERDATICS, Shaping the future: emerging technologies in renewable energy projects, July 2023. Accessed 20 April 2025
8. T.M. Gür, Review of electrical energy storage technologies, materials and systems: challenges and prospects for large-scale grid storage. Energy Environ. Sci. **11**, 2696–2767 (2018)
9. R. Martin, Renewable energy policy initiatives in the united states: a review of state renewable portfolio standards. J. Energy Policy **25**, 124–139 (2020)
10. IRENA, *Renewable power generation costs in 2019.* International Renewable Energy Agency (2020)
11. R. Wiser, M. Bolinger, Annual technology baseline: wind power cost reduction. National Renewable Energy Laboratory (NREL) (2019)
12. K. Miller, S. Sorrell, Corporate climate action: the role of business in global decarbonization. Energy Policy **152**, 112211 (2021)
13. T.R. Ayodele, A.S.O. Ogunjuyigbe, O.A. Akinola, Wind energy resource analysis for power generation in selected sites in south-west Nigeria. Renew Energy **114**, 1367–1385 (2018)
14. S. Mekhilef, R. Saidur, A. Safari, A review on solar energy use in industries. Renew. Sustain. Energy Rev. **15**, 1777–1790 (2012)
15. M. Banja, F. Monforti-Ferrario, I.P. Pascua, Assessing the potential of floating solar photovoltaics in Europe. Renew. Energy **163**, 425–438 (2021)
16. P. Vithayasrichareon, J. Riesz, I. MacGill, Using renewables to hedge against future electricity industry uncertainties—an Australian case study. Energy Policy **76**, 43–56 (2015)
17. F. Jiang, W. Zhu, J. Yang, A digital twin for wind turbine condition monitoring. Energy Rep. **7**, 754–764 (2021)

Chapter 2
The Science of Renewable Energy

2.1 Solar Energy Technology

Solar energy is one of the most abundant, sustainable, and rapidly advancing sources of renewable energy. By capturing the power of sunlight, solar technologies offer a clean alternative to fossil fuels, significantly contributing to global efforts to reduce greenhouse gas emissions and combat climate change. Solar energy can be harnessed in multiple ways, including thermal applications, concentrated solar power systems, and most notably, photovoltaic (PV) technology that directly converts sunlight into electricity.

The development and optimization of solar energy systems rely heavily on understanding the underlying physical principles that govern the conversion of solar radiation into usable energy. In the following sections, we delve into the science behind photovoltaic technology, examining the structure and operation of PV cells, the materials employed, and the processes that enable the generation of electric current from sunlight.

2.1.1 Photovoltaic Cell Mechanics

Photovoltaic (PV) cells generate electricity by harnessing the energy of sunlight through a process known as the photoelectric effect. The key to understanding this conversion lies in the interaction between light (photons) and the semiconductor materials in the PV cell, which leads to the generation of an electric current. Below is a detailed breakdown of the mechanics.

D. Gupta et al., *The Roadmap to Renewable Energy*,
https://doi.org/10.1007/978-3-032-09609-8_2

2.1.1.1 The Structure of a PV Cell

A typical PV cell is made up of layers of semiconductor materials, usually silicon, that are designed to create an electric field when exposed to light. These layers consist of:

N-type Semiconductor Layer: This layer is doped with elements such as phosphorus, which adds extra electrons (negatively charged particles), giving it a negative charge [1] (Fig. 2.1).

P-type Semiconductor Layer: This layer is doped with elements like boron, which have fewer electrons than silicon. This creates "holes" (positively charged spaces where an electron could go) in the material, resulting in a positive charge [1] (Fig. 2.1).

These two layers form a p-n junction, where the N-type layer has an excess of electrons, and the P-type layer has an excess of holes. This creates an electric field at the junction, which is crucial for moving electrons [1].

Absorption of Sunlight (Photons)

When sunlight (which is made up of photons) strikes the photovoltaic cell, the photons are absorbed by the semiconductor material. Each photon carries energy, and when a photon hits an electron in the silicon atom, it transfers its energy to the electron, knocking it free from its position [2] (Fig. 2.1).

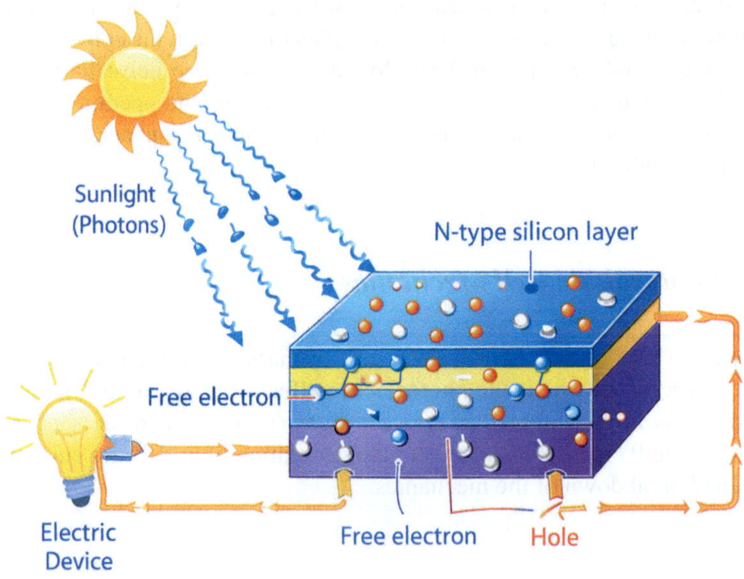

Fig. 2.1 Working principle of a photovoltaic solar cell (author-created)

This process is known as excitation. The energy of the photon must be greater than the bandgap energy of the semiconductor material for the electron to be freed. The silicon bandgap is ideal for absorbing a significant portion of the energy of sunlight [1].

Generation of Electron-Hole Pairs
Once a photon dislodges an electron, it leaves a hole (a positively charged spot) in the semiconductor material. The freed electron and the corresponding hole are known as an **electron-hole pair** [3] (Fig. 2.1).

In the presence of the electric field created by the p-n junction, the electron is pushed towards the n-type layer (negative side), while the hole moves towards the p-type layer (positive side). This separation of charges creates the conditions necessary for an electrical current to flow [2] (Fig. 2.1).

Flow of Electric Current
The electric field at the p-n junction drives the electrons towards the conductive front contact (metal grid on the surface of the cell), while the holes move toward the back contact of the cell. When the PV cell is connected to an external circuit, the electrons flow through the circuit, creating **direct current (DC)** electricity [3].

The movement of electrons through the external circuit provides usable electrical energy that can power devices or be stored in batteries. The flow of current continues as long as the PV cell is exposed to sunlight, providing a continuous source of electricity [2].

Conversion to Alternating Current (AC)
Since PV cells produce DC electricity, and most homes and businesses use alternating current (AC) electricity, an inverter is required to convert DC into AC. This allows the electricity generated by solar panels to be used for regular appliances or fed into the electrical grid [4].

Key Factors Affecting Efficiency
Bandgap Energy: The bandgap of the semiconductor material determines how much energy from the sunlight is converted into electricity. Silicon has a suitable bandgap that makes it highly efficient at capturing photons [3].

Sunlight Intensity and Angle: The amount of sunlight a PV cell receives directly affects its efficiency. PV cells perform best when sunlight strikes them perpendicularly, which is why tracking systems are sometimes used to optimize panel orientation [4].

2.1.2 Solar Thermal Technology

Solar thermal technology represents a pivotal approach to harnessing the Sun's abundant energy for practical applications, particularly electricity generation. Unlike photovoltaic (PV) systems, which directly convert sunlight into electricity, solar thermal systems capture and concentrate solar energy to generate heat, subsequently used to produce electricity or fulfill industrial thermal requirements. This section explores the principles of Concentrated Solar Power (CSP), its technologies, and diverse applications, emphasizing the innovative use of mirrors and lenses in focusing sunlight.

2.1.2.1 What Is Concentrated Solar Power (CSP)?

Concentrated Solar Power (CSP) is a solar thermal technology that uses optical devices like mirrors and lenses to concentrate sunlight onto a small, highly efficient receiver. The focused sunlight generates high temperatures, which are then used to heat a fluid, often molten salts or synthetic oils. This thermal energy can produce steam, driving a turbine connected to a generator to produce electricity.

2.1.2.2 Core Components of CSP Systems

1. **Solar Collectors**:
   - **Mirrors**: Key optical components that reflect and direct sunlight to a focal point or line.
   - **Lenses**: Occasionally used in advanced designs to focus sunlight onto receivers.
2. **Receivers**: Devices that absorb concentrated solar energy and transfer it to a working fluid.
3. **Thermal Energy Storage (TES)**: Allows for storing excess heat for use during non-sunny periods, ensuring electricity generation continuity.
4. **Power Block**: Includes turbines, generators, and associated equipment to convert heat into mechanical and then electrical energy.

2.1.2.3 Types of CSP Technologies

1. **Parabolic Trough Systems**:
   - Parabolic mirrors focus sunlight onto a linear receiver pipe at the focal point. (Fig. 2.2)
   - Heat transfer fluid flows through the pipe, absorbing solar heat for steam generation.
   - Commonly used due to their simplicity and cost-effectiveness.

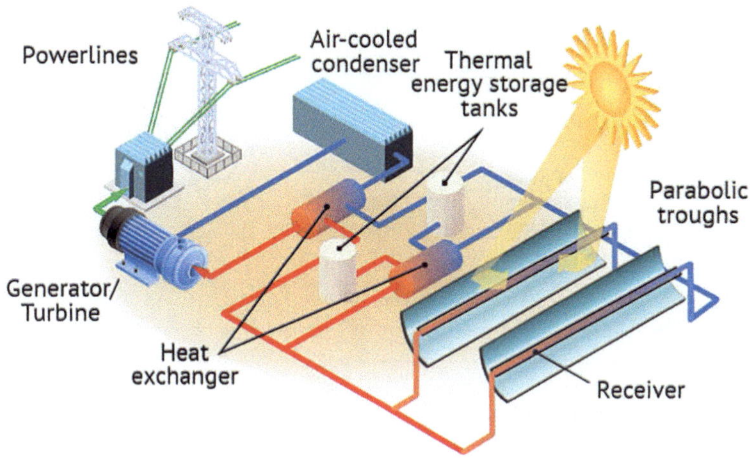

Fig. 2.2 Parabolic trough systems [5]

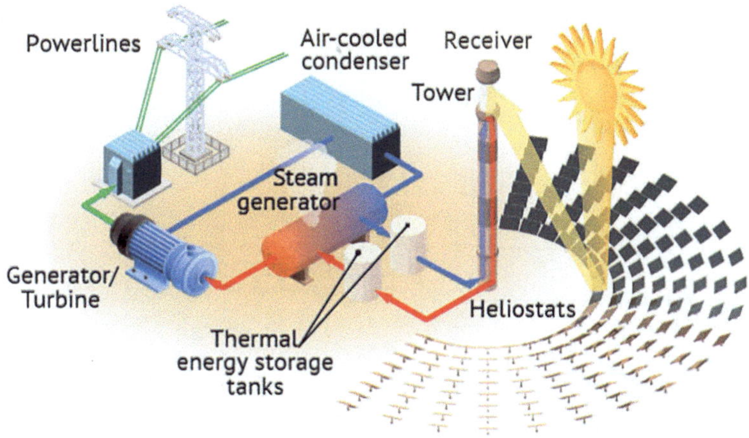

Fig. 2.3 Solar power towers [5]

2. **Solar Power Towers**:
   - Heliostats (flat, sun-tracking mirrors) concentrate sunlight onto a central receiver mounted on a tall tower (Fig. 2.3).
   - The receiver heats a working fluid, often molten salt, which acts as both a heat transfer medium and a storage medium.
3. **Dish/Engine Systems**:
   - Parabolic dish-shaped mirrors focus sunlight onto a receiver located at the focal point of the dish (Fig. 2.4).
   - The receiver transfers heat to a Stirling engine, directly generating electricity.

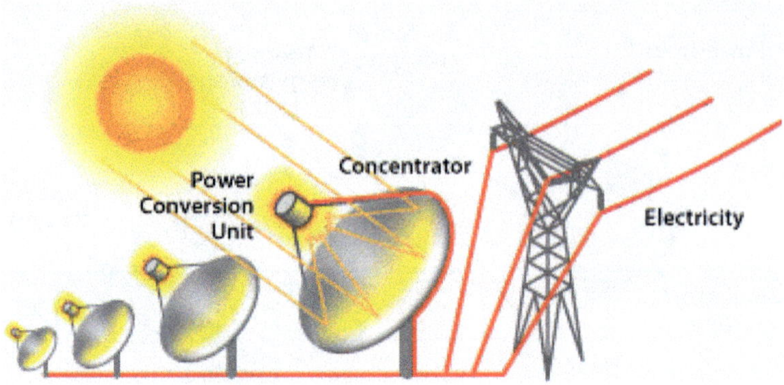

Fig. 2.4 Dish/engine receiver [6]

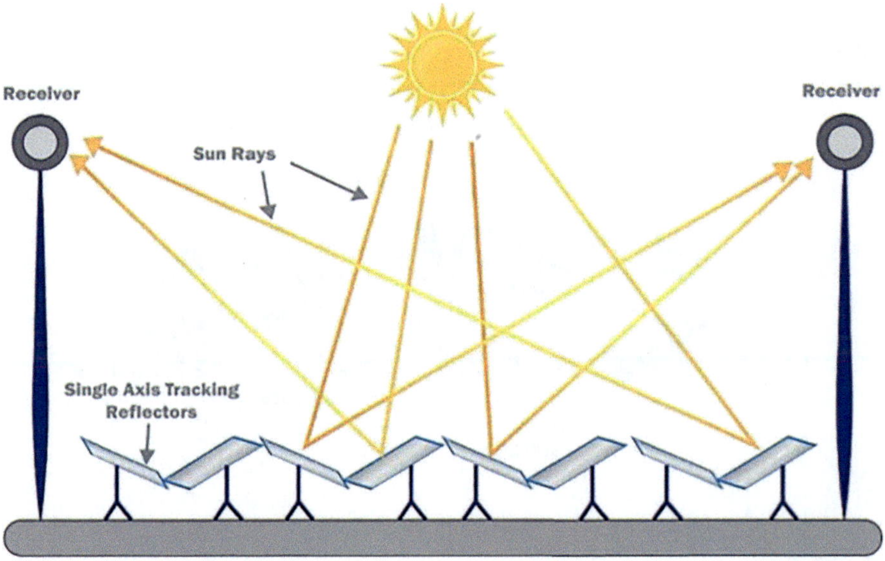

Fig. 2.5 Linear fresnel reflectors [7]

4. **Linear Fresnel Reflectors**:
   - Use flat or slightly curved mirrors to reflect sunlight onto a receiver located above the mirrors (Fig. 2.5).
   - Lower-cost alternative to parabolic troughs, suitable for smaller installations.

2.1.2.4 Applications of CSP

- **Utility-Scale Electricity Generation**: CSP plants are ideal for large-scale power generation, particularly in regions with high solar insolation, such as deserts and arid zones [8].

- **Industrial Process Heating**: CSP systems provide high-temperature heat for industrial applications, including food processing, chemical production, and water desalination [9].
- **Thermal Energy Storage**: Integrated CSP systems with molten salt storage enable grid stability by providing electricity during cloudy periods or at night [10].
- **Hybrid Systems**: CSP can be combined with other energy technologies, such as natural gas or biomass, to create hybrid systems for enhanced reliability [11].

2.1.2.5 Advantages of Using Mirrors and Lenses in CSP

- **Efficiency**: Concentrating sunlight increases thermal efficiency, maximizing energy capture [12].
- **Scalability**: CSP systems can be scaled from small industrial setups to vast utility plants.
- **Sustainability**: As a renewable energy source, CSP reduces greenhouse gas emissions and dependency on fossil fuels [13].
- **Storage Capabilities**: With thermal energy storage, CSP systems can generate electricity round-the-clock.

2.1.2.6 Challenges and Future Directions

Despite its benefits, CSP faces challenges, such as:

- **High Initial Costs**: The capital investment for CSP infrastructure is significant [14].
- **Land Use**: Large installations require considerable space, impacting land availability.
- **Geographic Limitations**: CSP systems perform best in regions with consistent, intense sunlight.

To address these challenges, researchers are focusing on advancements in mirror and lens materials, enhanced thermal storage solutions, and hybrid systems that integrate CSP with PV and other renewable technologies.

Solar thermal technology, particularly CSP, exemplifies the potential of concentrated sunlight as a sustainable energy source. By leveraging mirrors and lenses to focus sunlight, CSP systems efficiently generate electricity and meet diverse thermal energy demands. With continuous innovation, CSP is poised to play a critical role in the global transition to renewable energy, delivering clean, reliable power for a sustainable future.

2.1.3 *Solar Resource Assessment*

The efficient deployment of solar energy systems begins with a comprehensive understanding of the solar resource available in a given location. Solar resource assessment involves determining the potential of a geographic area to generate solar energy, taking into account factors such as solar irradiance, climate conditions, and geographic features. The accuracy of these assessments is critical for optimizing the design, financing, and operation of solar power plants, ensuring the highest energy yield and economic viability.

2.1.3.1 Techniques for Assessing Solar Potential

1. **Solar Irradiance**

   Solar irradiance refers to the power per unit area received from the Sun in the form of electromagnetic radiation. It is typically measured in watts per square meter (W/m^2). Understanding the distribution of solar irradiance over a given area is fundamental for evaluating its solar potential. The primary components of solar irradiance include:

   - **Global Horizontal Irradiance (GHI):** This measures the total solar radiation received on a horizontal surface, including both direct and diffuse radiation. GHI is a key metric for determining the overall solar potential, especially for flat-plate photovoltaic (PV) systems (Fig. 2.6) [15].
   - **Direct Normal Irradiance (DNI):** This measures the solar radiation received directly from the Sun on a surface perpendicular to the Sun's rays. DNI is important for concentrating solar power (CSP) systems, which require direct sunlight for effective energy generation (Fig. 2.6) [16].
   - **Diffuse Horizontal Irradiance (DHI):** This represents the radiation that reaches the surface after being scattered by the atmosphere. DHI is particularly relevant in areas with frequent cloud cover or haze (Fig. 2.6) [17].

   To assess solar irradiance, meteorological data is collected using ground-based solar measurement stations, satellite data, or remote sensing techniques.
2. **Geographic Factors**

   The geography of a region plays a significant role in determining solar resource availability. Several factors must be considered:

   - **Latitude:** The latitude of a location directly influences the angle at which sunlight strikes the Earth. Locations closer to the equator typically receive more direct sunlight year-round, making them ideal for solar energy production. (Fig. 2.7) [15]
   - **Altitude:** Higher altitudes can lead to higher solar irradiance due to the thinner atmosphere at greater heights, resulting in less scattering of sunlight. However, the influence of altitude on solar energy is generally more relevant in mountainous regions (Fig. 2.7) [16].

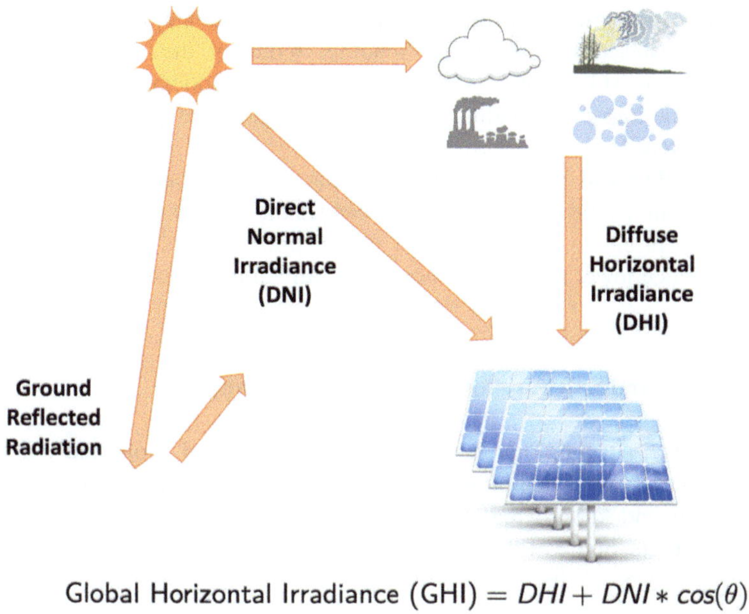

Fig. 2.6 Solar irradiance [18]

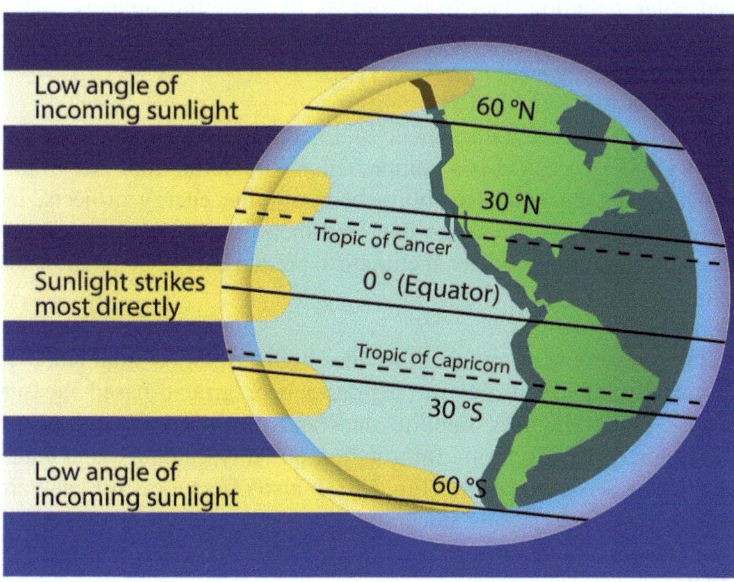

Fig. 2.7 Sunlight at different latitudes and altitudes [19]

- **Topography and Shading:** The presence of mountains, tall buildings, or other obstructions can create shading, reducing the amount of sunlight available to solar panels. Conducting a detailed topographic analysis helps assess the impact of shading on solar energy potential, particularly in hilly or urban areas [17].
- **Seasonality and Weather Patterns:** Seasonal variations in solar radiation are significant, as the angle of the Sun changes throughout the year. Additionally, regions prone to frequent cloud cover, precipitation, or dust storms will experience lower solar irradiance levels, affecting the overall solar potential.

2.1.3.2 Tools and Software for Solar Resource Mapping

Advancements in technology have provided a variety of tools and software to facilitate solar resource mapping and analysis. These tools use both ground-based measurements and satellite data to create accurate models of solar energy potential across large areas. Some of the most commonly used tools and software include:

1. **PVSyst**
   PVSyst is a popular software used for the simulation of photovoltaic systems. It allows users to model solar power plants, taking into account various factors such as location, weather conditions, system design, and shading. PVSyst uses detailed solar radiation data from sources like satellite imagery and ground-based measurements to provide an in-depth analysis of energy yield, system performance, and financial feasibility [20].
2. **SAM (System Advisor Model)**
   Developed by the U.S. Department of Energy's National Renewable Energy Laboratory (NREL), SAM is a comprehensive modeling tool that evaluates the performance and financial feasibility of renewable energy projects, including solar. SAM uses solar irradiance data and incorporates factors such as system losses, weather patterns, and local climate to predict energy generation and assess economic returns [21].
3. **SolarGIS**
   SolarGIS is a web-based tool that provides high-resolution solar resource maps, which are particularly useful for regions where ground-based measurements are scarce. SolarGIS uses satellite data and advanced algorithms to create solar irradiance maps, offering information on daily, monthly, and annual solar resource levels across the globe (Fig. 2.8). It also offers tools for analyzing solar potential at specific locations, integrating weather forecasts, and assessing long-term performance [22].
4. **SOLPOS**
   SOLPOS (Solar Position and Intensity) is an algorithm developed by NREL to determine the position of the Sun in the sky at any given time and location. It is widely used to calculate solar irradiance and assess the optimal orientation and tilt of solar panels for maximum energy generation [24].

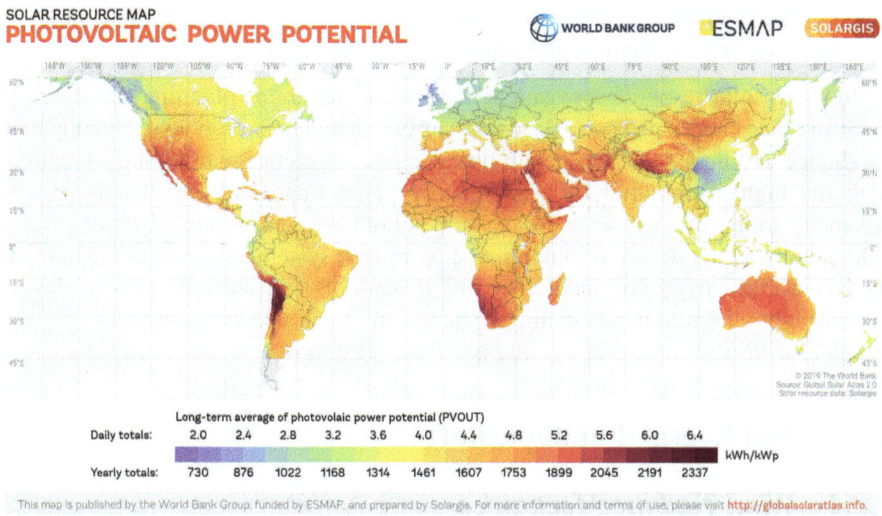

Fig. 2.8 Photovoltaic power potential [23]

5. **ArcGIS with Solar Radiation Tools**
   ArcGIS, a Geographic Information System (GIS) software, offers a suite of solar radiation analysis tools. These tools enable users to calculate solar radiation at a given location, considering factors such as latitude, altitude, and topography. ArcGIS is used in conjunction with digital elevation models (DEMs) to perform shading analysis, which is essential for identifying optimal sites for solar panel installation and assessing the impact of topography on solar resource availability [25].
6. **HOMER**
   HOMER (Hybrid Optimization of Multiple Energy Resources) is a software tool used for the design of distributed energy systems, including solar power. It models and optimizes hybrid renewable energy systems and helps assess the economic viability of integrating solar energy with other power sources such as wind, storage, and diesel generation [26].
7. **HelioScope**
   HelioScope is a cloud-based software for designing and optimizing solar PV installations. It provides detailed modeling of solar energy production using high-quality irradiance data, shading analysis, and system layout tools. HelioScope's user-friendly interface and advanced analytical capabilities make it an essential tool for solar professionals [27].
8. **Meteonorm**
   Meteonorm is a software tool that provides global solar and meteorological data, which is vital for the performance assessment of solar systems. It offers data

on solar irradiance, temperature, and other climatic conditions, and helps in simulating solar system performance for different locations worldwide [28].

In short, solar resource assessment is a crucial step in the planning and development of solar energy projects. By using various techniques such as analyzing solar irradiance and understanding geographic factors, developers can identify locations with the highest potential for solar energy generation. Additionally, the use of advanced tools and software for solar resource mapping enables more precise calculations, better decision-making, and optimized system designs. As the demand for renewable energy continues to grow, these tools and techniques will play an increasingly vital role in advancing the adoption of solar power globally.

2.2 Wind Energy Fundamentals

2.2.1 Wind Turbine Mechanics

Wind turbines are sophisticated machines designed to capture the kinetic energy of moving air and convert it into electricity. They operate based on fundamental mechanical and aerodynamic principles.

The major components of a modern horizontal-axis wind turbine include (Fig. 2.9):

- **Blades**: Typically three, shaped aerodynamically to maximize lift and minimize drag.
- **Rotor Hub**: Connects blades and transfers their rotational energy.
- **Nacelle**: Houses the gearbox, generator, and brake systems.
- **Tower**: Elevates the rotor and nacelle to heights where wind speeds are higher and turbulence is lower.

The blades function like airplane wings, utilizing the aerodynamic principle of lift. As wind flows over the curved surface of the blades, the air pressure decreases on one side, causing the blades to rotate around the hub. This rotational mechanical energy drives the main shaft, which is connected through a gearbox (in most designs) to a generator that produces electricity [30].

Notably, direct-drive turbines eliminate the gearbox to reduce maintenance costs and improve reliability. The power captured from the wind is proportional to the **cube of the wind speed** [31]. Therefore, even small increases in wind speed dramatically increase the energy output, making site selection critical.

The efficiency of a turbine is limited by the **Betz limit**, which states that no turbine can capture more than 59.3% of the kinetic energy in the wind [32].

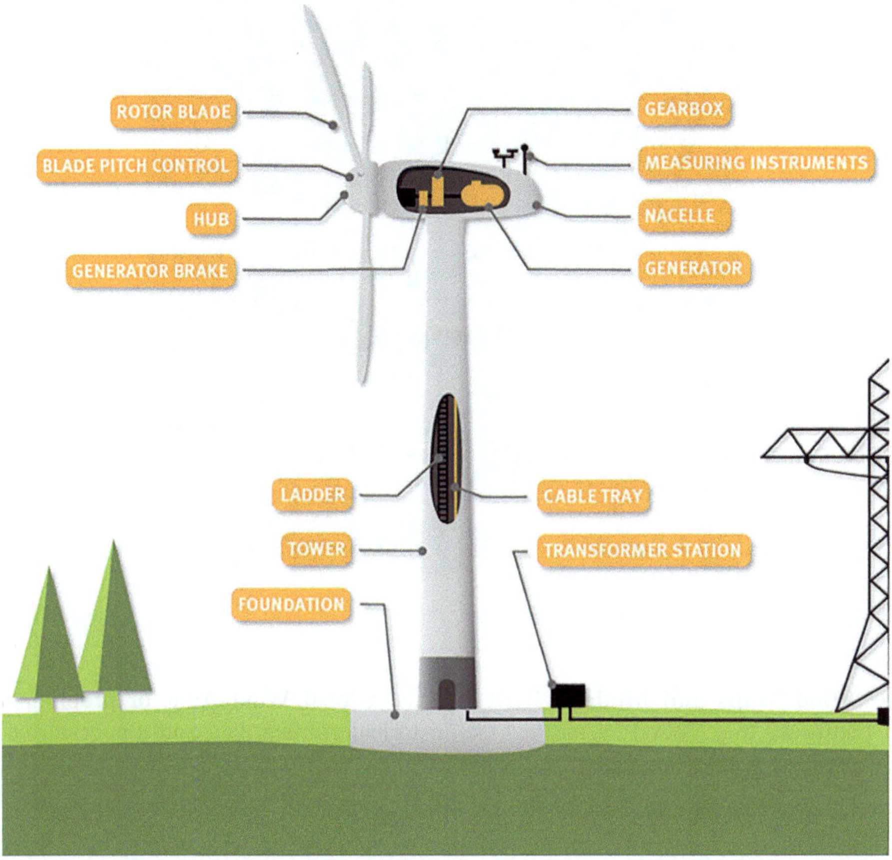

Fig. 2.9 Components of wind turbine [29]

2.2.2 *Wind Resource Assessment*

Accurately assessing wind resources is crucial to the technical and financial success of wind energy projects. Developers deploy a combination of short-term field measurements and long-term meteorological data modeling to characterize a site's wind regime.

Key steps in wind resource assessment include:

- **On-site Measurements**: Meteorological towers equipped with anemometers and wind vanes (Fig. 2.10) are installed at different heights (typically 60–120 m) to measure wind speed and direction over at least one year.
- **Remote Sensing**: Technologies like LiDAR and SoDAR allow for vertical wind profiling without tall towers [33].

Fig. 2.10 Anemometer and wind vane

- **Wind Maps and Models**: Tools like the Global Wind Atlas or proprietary software (e.g., WAsP, WindPro) use historical climate data and topography to estimate wind resources.

Wind speed variability is typically described statistically using a Weibull distribution. Turbulence intensity, shear (variation of wind speed with height), and wake effects from nearby turbines are also analyzed.

Locations with **average annual wind speeds exceeding 6–7 m per second** are generally considered viable [34]. Moreover, projects must evaluate seasonal patterns, extreme gusts, and icing risks, all of which impact turbine selection and maintenance strategies.

Economic feasibility hinges on ensuring a high **capacity factor**, the actual output over a year relative to maximum possible output.

2.2.3 Offshore Wind Energy

Offshore wind energy projects harness the stronger and more consistent winds over oceans and large lakes. Offshore wind farms generally achieve higher capacity factors (often exceeding 45–50%) compared to onshore farms (30–35%) [35].

However, offshore installations face unique engineering and logistical challenges:

- **Foundation Types**: Shallow water turbines use monopiles or jackets, while floating platforms are developed for deeper waters exceeding 50 m depth.
- **Corrosion Protection**: Marine environments are highly corrosive; thus, specialized coatings and cathodic protection systems are essential.
- **Grid Connection**: Subsea cables and offshore substations are necessary to transmit power to land.
- **Installation and Maintenance**: Heavy-lift vessels and helicopters are often required for installation and servicing, increasing costs.

Despite higher upfront costs, offshore projects benefit from:

- Higher wind speeds
- Minimal visual impact compared to land installations
- Larger available areas, enabling gigawatt-scale projects.

Europe leads offshore development, with projects like **Hornsea One** (UK) and **Borssele** (Netherlands) demonstrating the technology's potential [36].

More on offshore wind energy is explained in Chap. 5 of this book.

2.3 Battery Storage Systems

2.3.1 Types of Energy Storage

Energy storage technologies are essential for mitigating the intermittency of renewable energy sources like solar and wind. Various types of battery technologies are deployed depending on the specific grid and storage needs.

Lithium-Ion Batteries dominate today's energy storage market due to their high energy density, relatively long cycle life, and rapidly decreasing costs [37]. Their chemistry typically involves a lithium cobalt oxide ($LiCoO_2$) cathode and a graphite anode, although newer variants use materials like nickel manganese cobalt (NMC) or lithium iron phosphate (LFP) for improved safety and lifespan. Figure 2.11 shows the charge–discharge cycle of a lithium-ion battery.

Solid-State Batteries aim to replace the flammable liquid electrolyte with a solid material, enhancing safety and energy density. Although promising, these batteries face challenges in manufacturability and scaling [38]. Figure 2.12 shows a schematic view of a solid-state battery, illustrating its structure and charge flow.

Flow Batteries, such as vanadium redox flow batteries (Fig. 2.13), separate the energy storage medium from the power conversion device. This allows energy and power to be scaled independently, making them ideal for large, long-duration applications [39].

Each technology has its advantages:

- Lithium-ion: Best for residential, vehicle, and fast-response grid applications.
- Flow batteries: Best for multi-hour or daily storage needs.
- Solid-state batteries: Future promise for enhanced energy and safety.

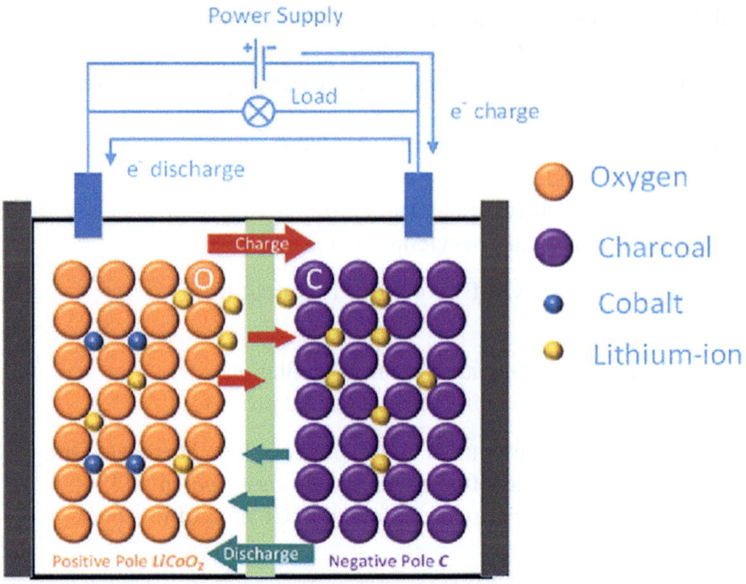

Fig. 2.11 Schematic view of a lithium-ion battery showing the charge–discharge cycle

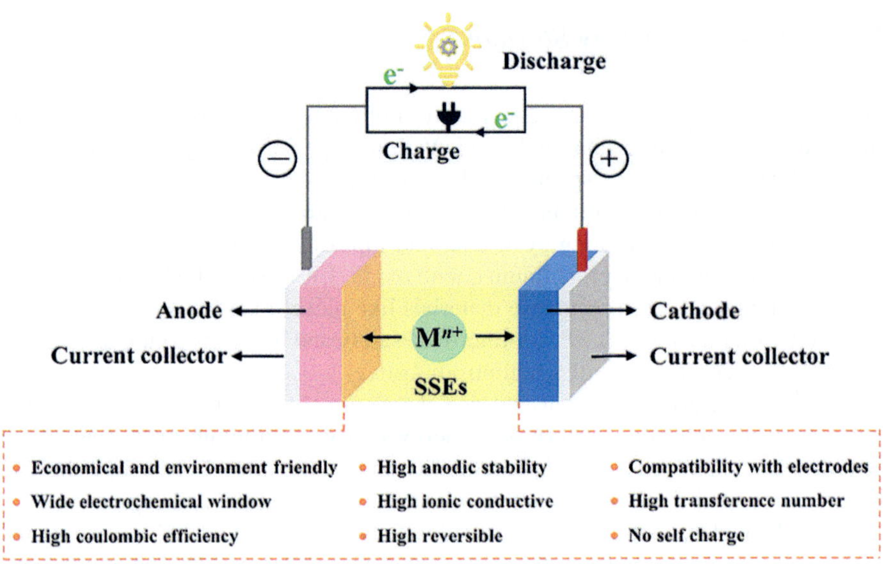

Fig. 2.12 Schematic view of a solid-state battery

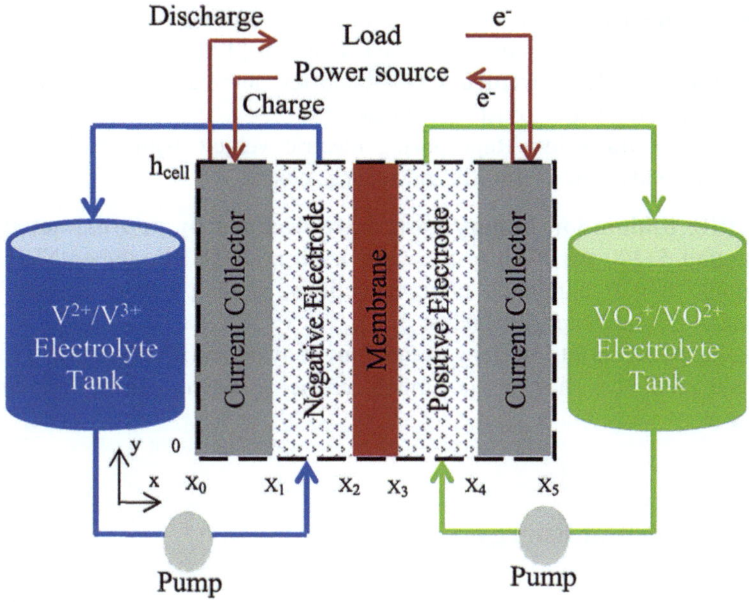

Fig. 2.13 Schematic view of a vanadium redox flow battery

2.3.2 Energy Storage Applications

Batteries provide multiple services that enhance grid resilience and efficiency:

- **Frequency Regulation**: Stabilizing grid frequency by absorbing or injecting small amounts of power on short timescales.
- **Peak Shaving**: Reducing peak demand charges by discharging during high-demand periods.
- **Renewable Integration**: Smoothing the variability of wind and solar generation.
- **Backup Power**: Providing power during outages for critical facilities like hospitals and data centers.

Hybrid systems like **solar-plus-storage** are increasingly common, allowing solar installations to store excess daytime generation for evening use [40]. Similarly, **wind-plus-storage** systems help mitigate the unpredictability of wind patterns.

Microgrids incorporating battery storage can operate independently of the central grid (island mode) during outages, enhancing resilience in disaster-prone areas [41].

2.3.3 Technological Advancements in Storage

Significant progress is underway to improve battery performance:

- **New Chemistries**: Lithium-sulfur and sodium-ion batteries offer the potential for lower-cost, more sustainable storage solutions [42].
- **Manufacturing Improvements**: Gigafactories, automation, and materials recycling are driving battery cost reductions.
- **Second-Life Batteries**: Repurposing electric vehicle batteries for stationary storage extends their useful life and reduces costs.

Since 2010, **lithium-ion battery costs have declined by approximately 89%**, from around $1200 per kWh to $132 per kWh in 2021 (BloombergNEF, 2021). Projections suggest continued price reductions, making battery storage a core component of future energy systems.

Ongoing research into **battery management systems (BMS)**, thermal management, and predictive analytics is further enhancing battery safety, longevity, and integration.

2.4 Geothermal Energy

Heat within the Earth originates from three sources: primordial heat from its formation, radioactive decay of elements, and frictional heating from mantle movements influenced partly by lunar tides. The core reaches temperatures above 5200 °C, hotter than the surface of the Sun, radiating about 47 TW of energy into space. This geothermal heat can be used for electricity generation, building heating, and temperature regulation for heat pumps, offering a significant renewable resource that remains underutilized.

2.4.1 Geothermal Heat Extraction

Successful geothermal power generation depends on three factors: shallow hot rocks, fluids to transport heat, and rock permeability. Ideal temperatures for live steam (200–400 °C) are more accessible in tectonically active areas like Kilauea (Hawai'i), Yellowstone, mid-ocean ridges, volcanic arcs such as Japan and the Aleutians, and at sites like Iceland, New Zealand, and The Geysers near Santa Rosa, California.

Natural hydrothermal systems require circulating groundwater, as seen at Yellowstone's Old Faithful geyser, where heated water turns to steam and erupts when pressure overcomes the overlying water's weight. Hot springs represent less violent versions of this phenomenon. Permeability, provided by fractures or interconnected pore spaces, is crucial for fluid flow but can be diminished by changes in underground stress, overpumping, or earthquakes.

Medium-temperature geothermal energy can be used for direct heating in "deep direct use" (DDU) systems, particularly in sedimentary basins or repurposed depleted oilfields. Lower-temperature geothermal energy supports shallow buried loops for residential heating and cooling, exploiting the constant 4.5–21 °C ground temperatures.

A challenge in high-temperature geothermal plants is mineral scaling. Dissolved minerals such as quartz and calcite precipitate when the high-pressure, high-temperature fluids reach the surface. Mammoth Hot Springs in Yellowstone showcases natural calcite terraces. In plants like Rotokawa in New Zealand, mineral buildup frequently clogs pipes; however, valuable minerals like gold offset maintenance costs.

2.4.2 Geothermal Resource Evaluation

Historically, geothermal development has been confined to hydrothermal sites—natural surface manifestations of deep Earth heat—located along tectonic boundaries. Consequently, geothermal energy provides less than half a percent of the U.S. energy supply, mainly along the west coast. In contrast, Iceland, positioned on the Mid-Atlantic Ridge, derives about a quarter of its electricity and heats 85% of its buildings from geothermal energy. Other geothermal-rich countries include Indonesia, New Zealand, and Kenya.

Geothermal resource evaluation hinges on understanding the geothermal gradient—the rate at which temperature increases with depth. Bottom hole temperature (BHT) measurements from deep wells, adjusted for surface air temperature (assumed 15 °C), are used to estimate gradients. Typical gradients average 25–30 °C per kilometer (15 °F per 1000 ft). To achieve the 200 °C necessary for steam generation, drilling depths of about 6 km are often required, although favorable areas in the western U.S., Gulf Coast, and Appalachians may reach adequate temperatures at 2–3 km.

However, BHT measurements can be distorted by fluid movement. Continuous temperature logging, commonly conducted by oil and gas companies, provides more reliable geothermal gradient profiles and is available through public records at state geological surveys.

New exploration focuses on "hidden geothermal systems"—hot groundwater aquifers without surface indicators. For example, the Black Hills of South Dakota recharge deep aquifers that heat up underground, producing wells with temperatures of 60–80 °C [43] farther east. The U.S. Department of Energy's Geothermal Technology Office (GTO) is applying "play fairway analysis" (PFA), a petroleum exploration technique, to locate hidden resources, successfully demonstrated in Nevada, Idaho, Washington, and Hawai'i.

GTO also promotes the use of geophysical imaging (gravity surveys, seismic lines, magnetotellurics) combined with geological and geochemical surface studies to model and characterize underground geothermal reservoirs.

2.4.3 *Types of Geothermal Power Plants*

Three conventional types of geothermal plants exploit hydrothermal systems based on fluid temperature:

- **Direct Dry Steam:** High-temperature resources (200–400 °C) generate pure, particle-free steam that can directly spin turbines. Any entrained water droplets could damage turbine blades.
- **Flash and Double-Flash Steam:** Fluids above 182 °C, held under pressure, flash into steam when depressurized. Double-flash plants use the residual hot liquid to extract additional energy.
- **Binary Cycle Power:** Moderate-temperature fluids (below 200 °C) transfer heat to a secondary fluid with a lower boiling point (e.g., alcohol, hexane, HCFC refrigerants) in a closed Rankine cycle system. Though less powerful than dry steam, binary systems are more scalable and may dominate future geothermal power development.

References

1. L.M. Fraas, *Low-Cost Solar Electric Power* (Springer, Berlin, 2014)
2. C. Honsberg, S. Bowden, *Photovoltaics: A Guide to the Solar Power Revolution* (PV Education, 2023)
3. J. Nelson, *The Physics of Solar Cells* (Imperial College Press, London, 2003)
4. R.A. Messenger, J. Ventre, *Photovoltaic Systems Engineering* (CRC Press, Boca Raton, 2010)
5. Nickel Institute, Renewable energy shines with concentrated solar power (2022)
6. U.S. Department of Energy. Dish/engine system: Concentrating solar-thermal power basics. https://energy.gov/eere/solar/dishengine-system-concentrating-solar-thermal-power-basics (n.d.). Office of Energy Efficiency & Renewable Energy. Retrieved 23 Dec 2024
7. D.R. Mills, G.L. Morrison, Compact linear fresnel reflector solar thermal powerplants. Solar Energy **68**(3), 263–283 (2000)
8. U.S. Department of Energy, Concentrating solar-thermal power basics. https://www.energy.gov/eere/solar/concentrating-solar-thermal-power-basics (n.d.). Accessed 27 April 2025
9. SolarPACES, How concentrated solar power works. https://www.solarpaces.org/worldwide-csp/how-concentrated-solar-power-works/ (n.d.). Accessed 27 April 2025
10. Southwest Research Institute, Concentrated solar power (csp) energy storage. https://www.swri.org/markets/energy-environment/power-generation-utilities/advanced-power-systems/concentrated-solar-power-csp-energy-storage (n.d.) Accessed 27 April 2025
11. ScienceDirect, Concentrating solar-thermal power technologies: status and analysis. *ScienceDirect* (2023). Accessed 27 April 2025
12. TTnergy, Concentrated solar power efficiency: using the power of the sun. https://www.ttnergy.com/concentrated-solar-power-efficiency-using-the-power-of-the-sun/ (2023). Accessed 27 April 2025
13. HeliosCSP, Concentrated solar power pros and cons. https://helioscsp.com/concentrated-solar-power-pros-and-cons/ (n.d.). Accessed 27 April 2025
14. ScienceDirect, The economics of concentrating solar power (csp): Assessing cost and performance factors. *ScienceDirect* (2023). Accessed 27 April 2025
15. Solar Anywhere, Definitions of solar data fields. https://www.solaranywhere.com/support/data-fields/definitions/ (2023). Accessed 28 April 2025

16. GML, Solar radiation measurements. https://gml.noaa.gov/grad/starmeas.html (2023). Accessed 28 April 2025
17. PVPMC, Global horizontal irradiance. https://pvpmc.sandia.gov/modeling-guide/1-weather-design-inputs/irradiance-insolation/global-horizontal-irradiance/ (2023) Accessed 28 April 2025
18. I. Zeroual, Equation of calculating ghi using dni and dhi (2022). Accessed 28 April 2025
19. Berkeley University of California, Reflection and absorption of sunlight. https://ugc.berkeley.edu/background-content/reflection-absorption-sunlight/ (2023). Accessed 28 April 2025
20. PVSyst, Pvsyst - photovoltaic simulation software. https://www.pvsyst.com (2025). Accessed 28 April 2025
21. National Renewable Energy Laboratory (NREL), System advisor model (sam). https://sam.nrel.gov (2023). Accessed 28 April 2025
22. SolarGIS, Solargis - solar resource maps. https://solargis.info (2023). Accessed 28 April 2025
23. Penn State University, Solar energy potential and utilization (2025). Accessed 28 April 2025
24. National Renewable Energy Laboratory (NREL), Solpos - solar position and intensity model. https://www.nrel.gov/grid/solar-resource/solpos.html (2023). Accessed 28 April 2025
25. Esri, Arcgis solar radiation tools. https://www.esri.com/en-us/arcgis/products/arcgis-pro (2023). Accessed 28 April 2025
26. HOMER Energy, Homer - hybrid optimization of multiple energy resources. https://www.homerenergy.com (2023). Accessed 28 April 2025
27. HelioScope, Helioscope - solar pv design and optimization software. https://www.helioscope.com (2023). Accessed 28 April 2025
28. Meteonorm, Meteonorm - global solar and meteorological data. https://meteonorm.com (2023). Accessed 28 April 2025
29. S.M. Francis, Onshore wind turbine basics: parts, power generation, and more. Long International (2023). Accessed online
30. P. Gipe, *Wind Power: Renewable Energy for Home, Farm, and Business* (Chelsea Green Publishing, White River Junction, 2004)
31. T. Burton, D. Sharpe, N. Jenkins, E. Bossanyi, *Wind Energy Handbook* (John Wiley & Sons, Hoboken, 2011)
32. A. Betz, Das maximum der theoretisch möglichen ausnutzung des windes durch windmotoren. Zeitschrift für das gesamte Turbinenwesen (1920)
33. J.F. Manwell, J.G. McGowan, A.L. Rogers, *Wind Energy Explained: Theory, Design and Application* (John Wiley & Sons, Hoboken, 2009)
34. T. Ackermann, L. Söder, An overview of wind energy-status 2002. Renew. Sustain. Energy Rev. **6**(1–2), 67–127 (2002)
35. W. Musial, B. Ram, Large-scale offshore wind power in the united states: assessment of opportunities and barriers. Technical report, National Renewable Energy Laboratory (NREL) (2010)
36. B. Snyder, M.J. Kaiser, Ecological and economic cost-benefit analysis of offshore wind energy. Renew. Energy **34**(6), 1567–1578 (2009)
37. J.M. Tarascon, M. Armand, Issues and challenges facing rechargeable lithium batteries. Nature **414**(6861), 359–367 (2001)
38. J. Janek, W.G. Zeier, A solid future for battery development. Nat. Energy **1**, 16141 (2016)
39. M. Skyllas-Kazacos, M.H. Chakrabarti, S.A. Hajimolana, F.S. Mjalli, M. Saleem, Progress in flow battery research and development. J. Electrochem. Soc. **158**(8), R55–R79 (2011)
40. X. Luo, J. Wang, M. Dooner, J. Clarke, Overview of current development in electrical energy storage technologies and the application potential in power systems. Renew. Sustain. Energy Rev. **42**, 393–431 (2015)
41. P. Denholm, M. Hand, Grid flexibility and storage required to achieve high penetration of variable renewable electricity. Energy Policy **39**(3), 1813–1820 (2011)
42. A. Manthiram, A reflection on lithium-ion battery cathode chemistry. Nat. Commun. **11**(1), 1550 (2020)
43. J.W. Lund, South dakota geothermal resources. Geo-Heat Center Q. Bull. **18**(4), 1–6 (1997)

Chapter 3
Policy and Regulatory Frameworks for Renewable Energy

3.1 Global and Regional Energy Policies

Renewable energy adoption is deeply influenced by policy and regulatory frameworks established at global, regional, and national levels. These frameworks are designed to promote investment in clean technologies, encourage research and development, and create market conditions favorable to the growth of renewable energy industries. International treaties and collaborations set the overarching goals for emission reductions and sustainability, while regional alliances and country-specific legislation translate these ambitions into actionable mandates and incentives.

Policy measures such as carbon pricing, renewable portfolio standards, feed-in tariffs, tax credits, and investment subsidies have been instrumental in accelerating the deployment of renewable energy technologies. Furthermore, initiatives that address energy access, energy security, and technological innovation reflect the varied priorities of developed and developing nations.

Understanding the evolution and impact of these frameworks is essential for evaluating the progress and challenges of the global renewable energy transformation. In the following sections, we examine major international climate agreements, national renewable energy standards, the role of regional collaboration in driving the renewable energy agenda forward, policy, trends and future direction.

3.1.1 International Climate Agreements

International climate agreements have become the foundation of renewable energy policies worldwide. Among these, the **Paris Agreement** (2015) is a landmark accord involving nearly every nation. It aims to limit global warming to well below 2 °C, preferably 1.5 °C, above pre-industrial levels (United Nations Frame-

D. Gupta et al., *The Roadmap to Renewable Energy*,
https://doi.org/10.1007/978-3-032-09609-8_3

work Convention on Climate Change [UNFCCC], 2015). The Agreement requires countries to develop and submit Nationally Determined Contributions (NDCs), which act as road maps for achieving their climate goals. These NDCs have catalyzed renewable energy adoption by prioritizing investments in wind, solar, and hydroelectric power while phasing out fossil fuels.

Additionally, initiatives such as the **Renewable Energy 100 (RE100) Campaign**, which encourages companies to commit to 100% renewable electricity, complement these agreements. For example, multinational corporations like Google and Apple have pledged carbon neutrality, reinforcing international climate goals [1].

Programs like **Sustainable Energy for All (SE4ALL)** and **Mission Innovation** foster cooperation among governments, businesses, and civil society to ensure universal access to affordable, reliable, and sustainable energy. By bridging the energy equity gap, these initiatives emphasize the global imperative to transition to renewable energy while supporting underdeveloped regions in overcoming technical and financial barriers [2].

3.1.2 National Renewable Energy Standards

On a national level, renewable energy standards and mandates serve as critical tools to meet climate goals while fostering economic development:

- **United States:** Renewable Portfolio Standards (RPS) exist in 30 states, requiring utilities to generate a specific percentage of their electricity from renewable sources. For instance, California's RPS targets 60% renewable energy by 2030 and 100% clean energy by 2045. This has spurred investments in solar farms and offshore wind projects, making the state a leader in clean energy innovation (National Renewable Energy Laboratory) [3].
- **European Union:** The Renewable Energy Directive (RED II) sets legally binding targets for EU member states to achieve at least 40% renewable energy in their energy mix by 2030. The EU Emissions Trading System (EU ETS) complements these targets by pricing carbon emissions, encouraging the adoption of low-carbon technologies [4].
- **India:** Through the National Solar Mission, India aims to achieve 280 GW of solar energy by 2030. Policies like mandatory renewable purchase obligations (RPOs) for utilities have facilitated the rapid expansion of solar parks and wind farms [5].

While developed nations often focus on advanced technologies, emerging economies emphasize scalable solutions like mini-grids and decentralized solar systems, addressing energy access challenges in rural areas. These tailored strategies highlight the diversity of approaches needed to achieve a global renewable energy transition.

3.1.3 State and Local Energy Policies

State and local governments often serve as trailblazers in renewable energy, setting policies and regulations tailored to regional needs and resources. These initiatives frequently go beyond national mandates, offering innovative approaches to clean energy adoption.

- **California:** The state's leadership in clean energy is evident through its aggressive targets, such as 100% clean electricity by 2045. The California Solar Initiative and programs supporting energy storage and electric vehicle infrastructure are accessible via the California Energy Commission and the California Public Utilities Commission websites.
- **New York:** Policies like the Climate Leadership and Community Protection Act (CLCPA) can be explored through the New York State Energy Research and Development Authority (NYSERDA). NYSERDA provides resources on renewable energy projects, funding opportunities, and state regulations.
- **Texas:** Though known for its oil and gas, Texas leads the U.S. in wind energy production. Information on local incentives and policies can be found through the Texas Comptroller's Office and the Public Utility Commission of Texas [6].
- **Other States:** Many states have databases that consolidate their energy policies. For instance, the *Database of State Incentives for Renewables & Efficiency (DSIRE)* is a comprehensive resource that provides detailed information about state, local, utility, and federal incentives and policies promoting renewable energy [7].

State and local energy policies are available on the following platforms:

1. **State Energy Office Websites:** These often include state-specific renewable energy goals, incentives, and legislative updates. Examples include:
    - California Energy Commission (CEC)
    - New York State Energy Research and Development Authority (NYSERDA)
2. **Local Government Websites:** Municipalities often outline their renewable energy projects and community programs.
3. **Public Utility Commissions (PUCs):** PUCs regulate utilities and provide updates on renewable energy standards and projects.
4. **National Databases:**
    - Database of State Incentives for Renewables & Efficiency (DSIRE): A centralized database with comprehensive information on renewable energy and efficiency policies across the U.S.
    - Energy Information Administration (EIA): Offers state energy profiles and data on renewable energy adoption.
5. **Non-Governmental Organizations (NGOs):** Organizations like the American Council for an Energy-Efficient Economy (ACEEE) provide reports and rankings of state energy policies.

These resources ensure that policymakers, businesses, and residents have access to up-to-date information on how to engage with and benefit from renewable energy programs.

3.2 Subsidies, Incentives, and Tax Credits

Subsidies, incentives, and tax credits are critical tools for fostering the growth of renewable energy by reducing financial barriers for developers, investors, and consumers. These mechanisms encourage investment, stimulate innovation, and promote the widespread adoption of clean energy technologies.

3.2.1 Federal Subsidies and Grants

Federal subsidies and grants are instrumental in enabling large-scale renewable energy projects, ensuring their economic viability and competitiveness with fossil fuels.

3.2.1.1 Investment Tax Credit (ITC)

The **Investment Tax Credit (ITC)** is a foundational component of U.S. federal renewable energy policy. Initially introduced under the *Energy Policy Act of 2005*, the ITC enables project developers to deduct a percentage of the capital cost of eligible renewable energy systems from their federal tax liability [8].

Under the *Inflation Reduction Act (IRA) of 2022*, the ITC was expanded, extended, and restructured to promote long-term clean energy investments:

- **Solar energy systems** are eligible for a **30% base ITC** through at least **2032**, with a possible extension until 2035 depending on national emissions reduction targets [9].
- Projects may qualify for **bonus credits**, potentially raising the total incentive up to **70%**, based on:
  - **10%** for meeting domestic content requirements (i.e., use of U.S.-manufactured materials),
  - **10%** for siting projects in designated energy communities,
  - **Up to 20%** for delivering benefits to low-income communities [10].

These enhancements under the IRA have significantly improved project economics, reduced investment risk, and driven workforce growth. Since the ITC's inception in 2006, the U.S. solar industry has grown by more than **10,000%** [11].

The clear, stable, and performance-based nature of the IRA's ITC structure has made it a central driver in accelerating the U.S. clean energy transition under the current administration.

3.2.1.2 Production Tax Credit (PTC)

The PTC incentivizes renewable energy production by providing financial support based on the electricity generated by eligible projects. Wind energy projects, for instance, can receive up to $0.026 per kilowatt-hour (kWh) for the first ten years of operation [12]. The PTC has been a driving force behind the U.S. becoming a global leader in wind energy, accounting for over 10% of total electricity generation.

3.2.1.3 Grants and Loan Programs

Federal agencies like the U.S. Department of Energy (DOE) offer grants and loan guarantees to support innovation and deployment of renewable technologies. Programs such as the DOE's Loan Programs Office (LPO) have provided funding for transformative projects, including utility-scale solar and advanced energy storage systems [8].

Federal support has been critical in reducing capital costs, fostering technological advancements, and increasing renewable energy deployment.

3.2.2 Incentives for Developers and Investors

Incentives tailored for developers and investors aim to enhance the financial attractiveness of renewable energy projects, creating a favorable environment for private sector participation.

1. **Feed-in Tariffs (FiTs):**
   FiTs guarantee a fixed, premium price for electricity generated by renewable sources over a specific period. These policies provide developers with predictable revenue streams, encouraging investment in renewable projects.
2. **Green Bonds:**
   Green bonds are financial instruments designed to raise capital for renewable energy and environmental projects. Issued by governments, corporations, or financial institutions, green bonds have mobilized billions of dollars for clean energy globally. For instance, in 2022, the green bond market exceeded $500 billion, with renewable energy projects comprising a significant share [13].

3. **Accelerated Depreciation:**
   Tax policies like the Modified Accelerated Cost Recovery System (MACRS) allow developers to recover capital costs of renewable energy assets over an accelerated schedule, improving project profitability [12].

These incentives reduce project risks and create a robust financial ecosystem to support renewable energy development.

3.2.3 Carbon Pricing and Market Mechanisms

Carbon pricing mechanisms provide economic incentives for reducing greenhouse gas (GHG) emissions, making renewable energy projects more competitive compared to fossil fuels.

1. **Cap-and-Trade Programs:**
   These programs set a limit on emissions and allow entities to trade allowances within this cap. California's cap-and-trade program has generated billions in revenue, which the state reinvests into renewable energy and energy efficiency initiatives [14].
2. **Carbon Taxes:**
   Carbon taxes directly price carbon emissions, encouraging businesses to adopt cleaner energy sources. Sweden's carbon tax, among the highest globally, has successfully driven a significant shift to renewable energy, reducing emissions by nearly 25% since its implementation in the 1990s [15].
3. **Voluntary Carbon Markets:**
   These markets enable businesses and individuals to purchase carbon offsets, funding renewable energy projects in developing countries. Initiatives like the Verified Carbon Standard (VCS) aim to ensure transparency and accountability in these transactions. However, the voluntary nature of these markets often results in a lack of standardization, leading to challenges in verifying the quality and impact of offsets across different programs [16].

Carbon pricing mechanisms align economic incentives with environmental goals, catalyzing a global transition to clean energy.

3.3 Regulatory Challenges

Renewable energy development faces numerous regulatory hurdles that can delay or even prevent projects from coming to fruition. These challenges span environmental permitting, grid integration, land-use zoning, and interconnection, all of which require nuanced navigation through complex legal frameworks and local conditions. In this section, we expand upon the critical regulatory challenges faced by renewable energy developers, providing further depth on the processes involved and strategies to overcome these barriers.

3.3.1 Environmental Permitting and Compliance

Environmental permitting and compliance represent some of the most significant regulatory challenges in renewable energy development. The permitting processes aim to ensure that projects meet environmental standards and minimize harm to ecosystems, communities, and natural resources. These regulations can vary considerably by country, state, and local jurisdiction, which further complicates the development process.

1. **National Environmental Policy Act (NEPA) and Other Federal Permitting**
   In the U.S., NEPA governs the environmental review process for projects that require federal funding or involve federal land. The act requires developers to assess and disclose the environmental impacts of their proposed projects, including impacts on air quality, wildlife, land use, and cultural resources. This is often a lengthy process that includes public comment periods, assessments, and potential litigation. For example, large-scale wind and solar projects on federal lands must undergo Environmental Impact Statements (EIS), which can take years to complete. These extensive reviews, while crucial for safeguarding the environment, can significantly delay renewable energy deployment [17].
2. **State-Level Environmental Permitting: CEQA and Other State Laws**
   - **California Environmental Quality Act (CEQA):**
     In states like California, CEQA mandates that projects undergo a thorough environmental review to assess potential impacts on natural resources, such as wildlife, water bodies, and air quality. This can lead to lengthy approval processes and opposition from environmental advocacy groups that seek to mitigate the project's environmental footprint. Developers are often required to implement mitigation strategies to reduce adverse impacts, such as wildlife protection measures for wind farms and restrictions on the use of water for solar thermal energy projects [18].
   - States with strong environmental protections, like New York and Oregon, also implement rigorous permitting processes, although they may offer streamlined reviews for renewable energy projects in some cases to incentivize clean energy adoption.
3. **Ecological Concerns:**
   - **Wildlife Protection and Habitat Fragmentation:**
     Wildlife conservation is a critical aspect of environmental reviews for renewable energy projects. Wind farms, in particular, have been subject to scrutiny regarding their impact on bird and bat populations. Some studies have shown that wind turbines can cause bird fatalities, particularly raptors and migratory

birds. Solar farms, meanwhile, can lead to habitat fragmentation, especially in areas of high biodiversity.
Developers must take precautions to mitigate these effects by conducting pre-construction environmental assessments and implementing wildlife monitoring programs during operation. In some cases, they may be required to modify project layouts, use bird-friendly turbine designs, or implement no-development buffers in sensitive areas [19].

- **Water Resource Management:**
  Renewable energy technologies such as geothermal, hydropower, and concentrated solar power (CSP) can have significant water usage implications. These technologies often require substantial amounts of water for cooling and steam production, potentially competing with local water needs, especially in arid regions. Developers must ensure that their projects meet local water management regulations and address any potential negative impacts on local water resources.

3.3.2 Grid Integration and Interconnection

As renewable energy adoption grows, one of the central challenges for developers and grid operators is integrating intermittent and geographically dispersed renewable resources into the grid. Regulatory frameworks around grid integration and interconnection are critical for facilitating a smooth transition to a renewable energy-dominant system. However, persistent uncertainty in these areas can also dampen investor confidence, which is essential for scaling project deployment.

1. **Interconnection and Regulatory Delays:**
   The process of connecting a renewable energy project to the grid involves a series of steps, including applying for interconnection, conducting feasibility studies, and obtaining necessary approvals from utility companies and grid operators. One of the primary challenges in this regard is the long wait times for interconnection approvals, particularly in areas where the grid is already oversubscribed. In the U.S., developers often face significant delays, sometimes years, before they can connect their projects to the grid, which affects their ability to start generating revenue. These delays introduce considerable financial risk, leading to growing investor concerns about the predictability of project timelines and returns. The Federal Energy Regulatory Commission (FERC)[1] has initiated efforts to streamline interconnection procedures, but much work remains to be done to address the backlog [20].

[1] FERC is an independent U.S. agency that regulates the interstate transmission of electricity, natural gas, and oil. It also oversees wholesale electricity markets, hydroelectric licensing, and ensures reliability of the electric grid.

2. **Transmission Planning and Investment:**
   Even if a renewable energy project successfully connects to the grid, challenges remain regarding the adequacy of transmission infrastructure. Many renewable energy resources, particularly wind and solar, are located in remote areas far from population centers. To efficiently transport electricity from these sites to demand areas, significant investments in new transmission infrastructure are required.

   - The planning and approval of new transmission lines can face opposition due to concerns about land use, environmental impacts, and community disruptions. Regulatory bodies such as FERC and Regional Transmission Organizations (RTOs) must navigate these challenges while balancing the needs of renewable energy developers and local stakeholders [8].
   - Some states, such as Texas, have undertaken efforts to develop transmission lines specifically designed to carry renewable energy from remote wind and solar sites to urban centers. However, these projects face significant regulatory hurdles, particularly when crossing state boundaries. These uncertainties in planning and permitting contribute to risk perception among investors, especially in markets where transmission development lags behind renewable project growth.

3. **Grid Stability and the Role of Energy Storage:**
   One of the main concerns about integrating renewable energy into the grid is the intermittency of sources like wind and solar. Fluctuations in energy production can cause grid instability, particularly during periods of low wind or cloudy days. Regulatory efforts to encourage the use of energy storage systems, such as battery storage, are critical to overcoming this challenge. Storage systems can store excess renewable energy generated during periods of high production and release it when demand is higher or when renewable generation is low. However, regulatory frameworks around energy storage are still evolving, and developers must navigate complex rules regarding compensation, performance standards, and integration with existing grid infrastructure [20]. The lack of clear and stable guidelines further contributes to investor hesitation, particularly in assessing the long-term viability and profitability of storage-integrated projects.

3.3.3 Land-Use and Zoning Regulations

Land-use and zoning regulations play a crucial role in determining where renewable energy projects can be built. These regulations vary significantly from region to region, creating barriers for developers who must navigate a patchwork of local, state, and national laws.

1. **Land-Use Conflicts and Zoning Laws:**
   Renewable energy projects often face opposition from local communities and landowners who are concerned about the impact of large-scale developments on

Fig. 3.1 Fountain Wind Project faced opposition [22]

agriculture, natural beauty, or property values. Zoning regulations are often used to restrict the types of land uses permitted in certain areas, which can make it difficult to site renewable energy projects, particularly wind farms and large-scale solar farms. In some cases, local opposition can lead to "Not In My Backyard" (NIMBY) sentiment, where communities oppose projects that, while beneficial at a global or state level, are seen as disruptive at a local level. For instance, in rural areas, solar farms may conflict with agricultural land use, while wind farms may face resistance due to aesthetic concerns and noise issues [19].

2. **Case Studies on Land-Use Conflicts:**

   - In **California**, the development of wind farms in rural areas often clashes with agricultural practices, leading to debates about land rights and land stewardship. Similarly, solar farms have faced opposition from local communities who view them as a threat to traditional farming or scenic vistas. Some states have passed laws to facilitate renewable energy projects, but local resistance and legal challenges remain significant obstacles.
     A notable case is the **Fountain Wind Project** in Shasta County proposed by ConnectGEN, where local officials rejected the proposal citing wildfire risks, interference with emergency services, and strong public opposition. Figure 3.1 shows an image circulated in the community expressing opposition to the Fountain Wind Project. This reflects how environmental and community safety considerations can override large-scale renewable ambitions [21].
     On the solar front, the **Ivanpah Solar** Electric Generating System in the Mojave Desert developed by BrightSource Energy with investment from NRG and Google, faced environmental pushback due to its impact on the endangered desert tortoise habitat. The project underwent legal and regulatory scrutiny, resulting in habitat conservation plans and mitigation efforts [23] (Fig. 3.2).

Fig. 3.2 Visual of Ivanpah Solar [24]

- **In New York**, a state-level initiative has sought to streamline zoning approvals for renewable energy projects by creating a comprehensive renewable energy siting process. The state's Office of Renewable Energy Siting (ORES) has the authority to approve renewable projects with a focus on minimizing land-use conflicts while advancing climate goals [25].

3. **Streamlining Zoning Regulations:**
   Several U.S. states and local governments are working on simplifying zoning regulations for renewable energy projects to overcome these challenges. Streamlining the permitting process through standardized zoning ordinances and clearer land-use policies is seen as a critical step in facilitating the deployment of renewable projects while ensuring community engagement and environmental protection.

3.4 Policy Trends and Future Directions

The future of renewable energy policy is being shaped by several emerging trends, which are designed to address the evolving energy landscape. These include the growing emphasis on decentralized energy solutions such as microgrids, the rise of green hydrogen as a clean energy source, and the ongoing development of carbon markets. In this section, we explore these key trends and their implications for future energy systems.

3.4.1 Decentralized Energy and Microgrids

The shift toward decentralized energy systems, where energy generation is distributed across a variety of small-scale, localized sources rather than centralized power plants, is gaining traction. Microgrids, in particular, are becoming a focal point of this trend, with a growing recognition of their potential to enhance energy security, resilience, and integration of renewable energy.

1. **The Rise of Decentralized Energy Solutions**
   Decentralized energy systems, which involve localized generation sources such as rooftop solar, wind turbines, and small-scale hydropower, have several advantages over traditional centralized energy grids. These systems reduce transmission losses, enable communities to generate and store their own energy, and contribute to energy independence. Furthermore, they facilitate the integration of renewable energy by allowing for a greater share of locally generated power to be consumed directly, minimizing the need for long-distance energy transmission.
   Policy Implications:
   Decentralized energy solutions require tailored policy frameworks that encourage local investment, ensure grid integration, and address regulatory barriers. Some governments are already moving towards recognizing the value of decentralized generation. For instance, countries like Germany and the United States have introduced policies that enable grid-connected residential solar systems, with net metering policies that allow homeowners to sell excess electricity back to the grid [8]. However, further advancements are necessary, particularly in addressing the regulatory frameworks for distributed generation and incentivizing investments in these technologies.
2. **Support for Microgrids**
   Microgrids are localized networks of electricity generation, storage, and distribution that can operate independently or in conjunction with the main grid. These systems are particularly valuable in regions with unreliable grid infrastructure, rural areas, or during extreme weather events, as they can continue to operate in isolation during grid outages.

   - **Regulatory Support**:
     Policymakers are increasingly recognizing the potential of microgrids for enhancing grid resilience and integrating renewable energy into local systems. The U.S. Department of Energy has supported the development of microgrids through funding programs and pilot projects, particularly those focused on integrating renewable energy technologies like solar, wind, and battery storage. Additionally, microgrids are supported by evolving regulations that

enable utilities and developers to design and implement these systems more easily.

- Some states, like **California** and **Hawaii**, have introduced policies and incentives aimed at accelerating microgrid development. These include financial incentives for deploying microgrid projects and simplified interconnection standards. As microgrids become more common, we can expect increased regulatory clarity and support, including policies that address issues such as grid compatibility, energy storage integration, and reliability standards [26].

3.4.2 Green Hydrogen and Emerging Technologies

Green hydrogen has emerged as one of the most promising solutions for decarbonizing hard-to-abate sectors, such as heavy industry and long-distance transportation. Produced through electrolysis powered by renewable energy, green hydrogen offers a clean alternative to fossil fuels. The potential for green hydrogen to revolutionize the energy landscape is being recognized by governments worldwide, leading to new policies that promote its development and integration into future energy systems.

1. **Policies Supporting Green Hydrogen Development**
   Governments are increasingly recognizing green hydrogen as a critical component of future energy systems. The European Union has been a leader in supporting the development of green hydrogen, launching the **EU Hydrogen Strategy** in 2020, which aims to establish hydrogen as a key enabler of a decarbonized economy. This strategy includes plans for infrastructure development, funding mechanisms, and a regulatory framework to accelerate green hydrogen production [4].
   U.S. Policies:
   In the United States, green hydrogen is gaining attention, with key policy initiatives such as the *Hydrogen Energy Earthshot* program, which seeks to reduce the cost of clean hydrogen production to $1 per kilogram. The *Bipartisan Infrastructure Law* also includes funding to support the development of regional hydrogen hubs and production facilities.
   However, it is important to note that not all hydrogen hubs are based solely on green hydrogen. Many of these hubs include a mix of hydrogen production methods. including **blue hydrogen** (produced from natural gas with carbon capture) and **pink hydrogen** (derived from nuclear energy),with the goal of building a broad-based hydrogen economy. These policies are designed to foster innovation, reduce costs, and make hydrogen a competitive low-carbon energy source in the long term [8].
2. **Regulatory Frameworks for Emerging Technologies**
   Alongside hydrogen, other emerging technologies such as advanced nuclear, small modular reactors (SMRs), and next-generation battery storage are also receiving attention from policymakers. As these technologies become increas-

ingly viable, regulatory frameworks will need to evolve to ensure they can be integrated into the energy mix.

- **Nuclear Fusion and Small Modular Reactors**:
  Policymakers are beginning to support the development of small modular reactors (SMRs), which offer the potential for safer, more cost-effective nuclear power. The U.S. Department of Energy has committed funding to research and development of SMRs and is working to streamline regulatory processes for their approval and deployment [8].
- **Advanced Battery Storage**:
  As energy storage technologies evolve, governments are creating frameworks to support large-scale deployment. For example, **California's Energy Storage Procurement Program** is designed to accelerate the deployment of energy storage systems, which will be critical for balancing the intermittent nature of renewable energy [27].

3.4.3 Future of Carbon Markets and Clean Energy Policy

As countries intensify their efforts to meet climate goals and reduce greenhouse gas emissions, carbon pricing mechanisms such as carbon taxes and cap-and-trade systems, are becoming increasingly central to clean energy policy. These tools provide economic incentives for reducing carbon emissions and promote the adoption of renewable energy technologies.

1. **Predictions on the Evolution of Carbon Pricing Mechanisms**
   Carbon pricing has become one of the most powerful tools in the fight against climate change. By placing a price on carbon, governments can create financial incentives for industries to reduce emissions, invest in clean energy, and adopt low-carbon technologies.

   - **Global Trends**:
     The **European Union Emissions Trading Scheme (EU ETS)** has been a pioneering initiative, setting a price on carbon emissions from the power, industrial, and aviation sectors. As of 2023, the EU is expanding its carbon market to cover additional sectors, and carbon prices are expected to rise, further incentivizing the transition to renewables. Other regions, including **California**, have implemented their own cap-and-trade systems, and countries like **Canada** have introduced carbon taxes to encourage emissions reductions [4].
   - **Future of Carbon Pricing**:
     Carbon pricing mechanisms are expected to expand globally, with countries implementing carbon markets and gradually increasing carbon prices. By 2030, many predictions indicate that carbon prices could reach levels that

significantly impact fossil fuel dependence, pushing more industries to adopt renewable energy and reduce their carbon footprints [15].

2. **Policy Outlook for Long-Term Climate and Renewable Energy Goals**
The policy landscape for renewable energy is rapidly evolving as governments worldwide strive to meet ambitious climate and energy goals. Key targets include net-zero emissions by 2050, as outlined by the **Paris Agreement**. To achieve these goals, continued policy innovation will be needed to support clean energy transitions, foster green hydrogen development, and expand energy storage capacities.
However, recent industry developments show some pullback and caution among investors and developers, reflecting uncertainties in technology costs, regulatory environments, and market dynamics. In response, the concept of climate "adaptation" is gaining prominence alongside mitigation strategies. Adaptation involves adjusting energy systems and infrastructure to withstand climate impacts and maintain reliability, recognizing that some level of change is inevitable even with aggressive emission reductions.

   - **Strategic Plans**:
   Governments are increasingly adopting **Nationally Determined Contributions (NDCs)** that outline their commitments to reducing emissions and transitioning to cleaner energy systems. These plans often include specific renewable energy targets, such as expanding solar and wind capacity or introducing mandatory renewable energy purchase obligations [28]. To remain effective, these strategies must now also integrate adaptation measures that enhance resilience and flexibility in the energy sector.
   - **Renewable Energy Integration**:
   Future policies will likely focus on accelerating renewable energy adoption through additional incentives, regulatory support for grid modernization, and market reforms that encourage investment in clean technologies. By aligning carbon pricing with renewable energy goals, policymakers can drive a smoother and more cost-effective transition to a low-carbon future. At the same time, regulatory frameworks will need to balance mitigation efforts with adaptation planning to address emerging climate risks and uncertainties.

References

1. Renewable Energy Policy Network for the 21st Century, Renewables global status report. https://www.ren21.net (2022)
2. International Energy Agency, World energy outlook 2021. https://www.iea.org (2021)
3. National Renewable Energy Laboratory (NREL), System advisor model (sam). https://sam.nrel.gov (2023). Accessed 28 April 2025
4. European Commission, Renewable energy directive (2020)
5. Ministry of New and Renewable Energy, National solar mission. https://mnre.gov.in (2023)

6. Texas Comptroller of Public Accounts, Wind energy snapshot – Texas energy report 2023. https://comptroller.texas.gov/economy/economic-data/energy/2023/wind-snap.php (2023). Accessed 28 April 2025
7. NC Clean Energy Technology Center, Database of state incentives for renewables & efficiency (dsire). https://www.dsireusa.org (2023)
8. U.S. Department of Energy, Transmission line planning and approval challenges (2023). Accessed 27 April 2023
9. Internal Revenue Service, Section 48 investment tax credit – overview and bonus credit criteria (2023)
10. Department of the Treasury, Inflation reduction act: guidance on bonus credits for energy communities and domestic content (2023)
11. Solar Energy Industries Association, Investment tax credit (itc). https://www.seia.org (2023)
12. Internal Revenue Service, Modified accelerated cost recovery system (macrs) - renewable energy, 2023. Accessed 27 April 2023
13. International Renewable Energy Agency, Green bonds for renewable energy financing. https://www.irena.org (2023)
14. California Air Resources Board, Cap-and-trade program overview. https://www.arb.ca.gov (2023)
15. World Bank, Carbon taxes and their impact on emissions. https://www.worldbank.org/en/news/press-release/2023/05/23/record-high-revenues-from-global-carbon-pricing-near-100-billion (2023)
16. Climate Action Reserve, Verified carbon standard (vcs) - voluntary carbon markets (2023). Accessed 27 April 2023
17. U.S. Department of Energy, Offshore wind energy: Block island case study. https://www.energy.gov (2023). Retrieved from DOE
18. California Governor's Office of Planning and Research, Ceqa guidelines for renewable energy projects. https://opr.ca.gov (2023)
19. National Renewable Energy Laboratory, State renewable portfolio standards. https://www.nrel.gov (2023)
20. Federal Energy Regulatory Commission, Grid integration of renewable energy resources. https://www.ferc.gov (2023).
21. Redding Record Searchlight, Shasta county rejects fountain wind project citing wildfire risk and public opposition. https://www.redding.com/story/news/2021/07/01/shasta-county-supervisors-reject-fountain-wind-project/7835686002/ (2021). Accessed 28 April 2025
22. Shasta County, California, Fountain wind project application deemed complete by cec. https://www.shastacounty.gov/community/page/fountain-wind-project-application-deemed-complete-cec (2024). Accessed 28 April 2025
23. U.S. Department of Energy, Ivanpah solar electric generating system. https://www.energy.gov/eere/solar/ivanpah-solar-electric-generating-system (2020). Accessed 28 April 2025
24. Google Earth and Pexels Contributor, An aerial shot of the ivanpah solar electric generating system. https://www.pexels.com/photo/an-aerial-shot-of-the-ivanpah-solar-electric-generating-system-14096756/ (2022). Accessed 28 April 2025
25. New York State Energy Research and Development Authority, Renewable energy programs. https://www.nyserda.ny.gov (2023)
26. California Public Utilities Commission, Microgrid support and policy. https://www.cpuc.ca.gov (2023)
27. California Public Utilities Commission, Community choice aggregation. https://www.cpuc.ca.gov (2023)
28. United Nations Framework Convention on Climate Change, Nationally determined contributions (ndcs). https://unfccc.int/ndc-synthesis-report-2023 (2023)

Chapter 4
Environmental and Social Considerations

4.1 Environmental Studies and Impacts

As renewable energy technologies continue to grow in scale and complexity, their interaction with the environment becomes increasingly significant. While these systems are designed to reduce carbon emissions and combat climate change, their implementation can still lead to environmental disturbances. For instance, large-scale solar farms may require land clearing that affects biodiversity, while wind turbines can pose risks to bird and bat populations. Hydropower projects may alter water ecosystems and disrupt local communities.

Environmental studies provide the framework to analyze, understand, and manage these interactions responsibly. By identifying potential adverse impacts early in the planning phase, developers can make informed decisions that align with sustainable development goals. These studies not only help in complying with legal and regulatory frameworks but also foster public trust and acceptance.

4.1.1 Role of EIAs in Renewable Energy Projects and Key Environmental Considerations

The renewable energy sector aims to deliver clean and sustainable energy solutions. However, the development of renewable energy infrastructure can sometimes conflict with environmental and social priorities. To address these concerns, Environmental Impact Assessments (EIAs) have emerged as a cornerstone in project planning and execution, ensuring that energy generation does not come at an unsustainable cost to the planet or local communities.

What is EIA? An Environmental Impact Assessment (EIA) is a structured and legally mandated process that evaluates the likely environmental effects of proposed development projects before they are implemented. It identifies, predicts,

D. Gupta et al., *The Roadmap to Renewable Energy*,
https://doi.org/10.1007/978-3-032-09609-8_4

and assesses potential environmental, social, and economic impacts while proposing strategies to mitigate negative outcomes.

The EIA process involves multiple stages, including screening, scoping, impact assessment, mitigation planning, public consultation, and decision-making. However, this process is expected to face significant setbacks over the next few years due to increasing regulatory scrutiny, evolving environmental standards, and potential resource constraints within review agencies. These challenges may lengthen approval timelines and add uncertainty for renewable energy projects, requiring developers to plan proactively to manage associated risks.

4.1.2 Importance of EIA in Renewable Energy Projects

Renewable energy projects, though designed to combat climate change, can have unintended environmental and social consequences. From habitat disruption to changes in local water systems, these projects demand a careful evaluation of their potential impacts. EIAs are essential in mitigating these risks and ensuring that renewable energy projects are both sustainable and socially responsible.

1. **Identifying Potential Risks:** Renewable energy projects often involve substantial land use, construction activities, and operational changes. For instance:
   - **Wind farms** may disturb wildlife habitats and migration patterns, particularly for birds and bats.
   - **Solar farms** can alter local land use, affecting agriculture, vegetation, and soil quality.
   - **Hydropower projects** may disrupt aquatic ecosystems and displace local communities. EIAs identify these risks early, enabling developers to plan for preventive measures.
2. **Promoting Compliance with Environmental Regulations:** Governments and international bodies have established stringent environmental laws and standards to protect ecosystems. EIAs help projects comply with these requirements by identifying areas of concern and aligning project designs with legal and regulatory frameworks.
3. **Enhancing Project Sustainability:** Sustainability lies at the heart of renewable energy initiatives. EIAs ensure that projects align with global sustainability goals by minimizing ecological harm and maximizing positive impacts. For example:
   - Designing wind turbines to avoid major bird migration routes.
   - Employing sustainable water management practices in solar thermal plants.
4. **Building Stakeholder Confidence:** A transparent EIA process fosters trust among stakeholders, including local communities, environmental groups, and regulators. By addressing concerns and demonstrating accountability, project developers can secure public support and limit resistance during project implementation.

4.1.3 Objectives of EIA

The objectives of EIAs provide a clear framework for managing environmental and social considerations. These objectives ensure that renewable energy projects proceed responsibly, with minimal harm to the environment and maximum benefits to society.

1. **Predicting Environmental Impacts:** EIAs assess the potential effects of a project on various environmental factors, including air quality, water resources, biodiversity, and land use. For example, a solar farm might require an assessment of how its large-scale land coverage affects local ecosystems and weather patterns.
2. **Providing Mitigation Measures:** Once potential risks are identified, EIAs recommend mitigation strategies to minimize negative impacts. Examples include:
   - Reforestation projects to offset deforestation caused by wind turbine installations.
   - Noise barriers for communities near wind farms.
   - Recycling or reusing water in hydropower operations.
3. **Ensuring Stakeholder Involvement:** Stakeholder involvement is vital to the success of renewable energy projects. EIAs integrate the voices of various stakeholders, such as local residents, indigenous groups, NGOs, and policymakers, into project decision-making. Effective consultation not only addresses public concerns but also leverages local knowledge to enhance project outcomes.
4. **Informing Decision-Makers:** Policymakers and regulatory authorities rely on EIA reports to make informed decisions regarding project approvals. These reports present comprehensive data, analyses, and recommendations, enabling decision-makers to weigh economic benefits against environmental and social costs.

4.1.4 Key Environmental Considerations in EIAs for Renewable Energy

In renewable energy development, specific environmental aspects require careful consideration during the EIA process:

1. **Biodiversity and Wildlife:** Renewable energy installations can affect local and migratory species. For example:
   - Wind turbines may pose collision risks to birds and bats.
   - Solar farms can disrupt habitats of terrestrial animals and pollinators.

   Mitigation measures might include setting aside conservation areas or employing wildlife-friendly designs.

2. **Water Resource Management:** Hydropower and solar thermal plants depend heavily on water resources. Overuse or contamination of local water systems can affect aquatic life and human communities. EIAs assess these risks and propose sustainable water management practices.
3. **Land Use and Visual Impact:** Large-scale energy projects often require significant land area, which can affect agriculture, forestry, and scenic landscapes. EIAs evaluate land use changes and visual impacts to minimize disruptions and preserve local aesthetics.
4. **Air and Noise Pollution:** While renewable energy systems generally reduce greenhouse gas emissions, their construction and operation can generate temporary pollution. For example, transporting materials for wind farms may increase vehicle emissions, and operational noise may affect nearby communities. EIAs address these concerns through strategic planning and technology choices.
5. **Cultural and Archaeological Impacts:** In some cases, renewable energy projects are located near culturally or archaeologically significant sites. EIAs ensure that these sites are preserved and respected, fostering cultural heritage protection.

4.2 Wetland Assessments

Wetlands are essential ecosystems that contribute significantly to biodiversity, water filtration, flood control, and carbon sequestration. In the context of renewable energy projects, these areas demand meticulous evaluation to ensure their preservation and to mitigate any potential environmental disruptions. Wetland assessments are thus integral to project planning and Environmental Impact Assessments (EIAs), ensuring compliance with environmental regulations and sustainable development goals.

4.2.1 Wetland Delineation

The delineation of wetlands involves identifying, mapping, and characterizing these ecosystems within or near a project area. This process is critical for understanding the spatial extent of wetlands and their ecological significance. Field surveys are a central component, where indicators such as hydrophytic vegetation, hydric soils, and wetland hydrology are observed and documented. Specialists often rely on tools like GIS and aerial imagery to supplement on-ground findings, creating detailed maps that inform project decisions. For instance, identifying high-value wetlands can lead to strategic adjustments in project design, such as shifting infrastructure away from sensitive zones [1] (United States Environmental Protection Agency [EPA], 2020).

The outcome of wetland delineation is a comprehensive report, including maps, data on ecological conditions, and evidence of compliance with environmental standards. This documentation not only guides developers in avoiding or minimizing wetland impacts but also forms a critical part of permit applications required under various regulatory frameworks [2].

4.2.2 Regulatory Framework for Wetlands

Wetland conservation is supported by a robust regulatory framework designed to prevent their degradation. The **Clean Water Act (CWA)** in the United States is a key piece of legislation, particularly Section 404, which governs activities that involve the discharge of materials into wetlands. Renewable energy developers must secure permits from the U.S. Army Corps of Engineers (USACE), demonstrating that the project avoids impacts where possible, minimizes unavoidable effects, and compensates for any remaining damage through mitigation measures [1].

Beyond federal law, many states enforce additional protections for wetlands, and international agreements like the Ramsar Convention emphasize the global significance of these habitats. The Ramsar Convention is an international treaty that provides a framework for the conservation and sustainable use of wetlands, and it is particularly significant for global wetland protection efforts [3]. Adherence to these legal standards ensures that projects align with conservation objectives while maintaining compliance with local and international norms.

4.2.3 Impact of Renewable Energy on Wetland Ecosystems

The expansion of renewable energy infrastructure poses distinct challenges to wetland ecosystems. While renewable projects aim to reduce overall environmental harm, their construction and operation can have localized effects on wetlands. The following outlines some of these effects and potential mitigation measures.

1. **Solar Energy Projects:** Large-scale solar developments may require substantial land clearing, which can disrupt natural hydrology, compact soil, and lead to habitat loss. However, mitigation measures such as siting projects outside wetland areas, employing elevated solar arrays, and using vegetation-friendly ground cover can significantly reduce these impacts [4].
2. **Wind Energy Projects:** Wind turbine installation can fragment wetland habitats due to the construction of turbines and access roads. Increased erosion and sedimentation can degrade water quality, impacting wetland health. Strategies to mitigate these impacts include maintaining buffer zones, using erosion control methods, and designing turbine layouts to minimize habitat disruption [5].

3. **Hydropower Projects:** Although hydropower is a renewable energy source, it can severely disrupt wetland ecosystems by altering natural flow patterns and inundating wetland areas. Thoughtful design and operation of these facilities, such as maintaining environmental flow regimes and employing fish passage mechanisms, are essential to mitigate such impacts [6].

Across all renewable energy types, general strategies for mitigating wetland impacts include prioritizing avoidance through careful site selection, minimizing unavoidable disturbances during construction, and engaging in compensatory restoration or wetland banking. Restoration involves repairing degraded wetlands, while wetland banking allows developers to compensate for impacts by funding conservation projects elsewhere [7].

4.2.4 Balancing Development and Conservation

The preservation of wetlands is vital for both environmental and societal benefits. Developers must strike a balance between advancing renewable energy projects and safeguarding these ecosystems. Proactive wetland assessments, adherence to regulatory frameworks, and implementation of thoughtful mitigation measures ensure that renewable energy development contributes to a sustainable future while maintaining the ecological integrity of wetland environments.

4.3 Wildlife and Habitat Assessment

Wildlife and habitat assessments are integral components of the Environmental Impact Assessment (EIA) process for renewable energy projects. These assessments ensure that the development of renewable energy infrastructure does not unduly harm wildlife populations or their habitats. Special attention is given to protected species, the potential impact of renewable energy systems on wildlife, and strategies to mitigate these effects.

4.3.1 Protected Species and Habitat Surveys

As part of the early-stage assessment process for renewable energy projects, protected species and habitat surveys are conducted to identify whether sensitive or protected species are present in the proposed project area. These surveys aim to locate the presence of species such as endangered birds, bats, amphibians, and aquatic species, as well as any critical habitats they may occupy. Protected species are often defined under local, national, or international regulations such as the

U.S. Endangered Species Act or the European Union Habitats Directive, which list species that are at risk of extinction.

Survey Methodology To assess the presence of protected species, a combination of field surveys, remote sensing, and acoustic monitoring is typically employed. Field surveys may involve direct observation of wildlife, tracking, and setting up camera traps to record animal activity. Remote sensing techniques, such as aerial imagery, can be used to identify large-scale habitat types, while acoustic monitoring is commonly used for detecting bat activity or migratory birds.

These surveys provide valuable data for understanding how a renewable energy project might affect local ecosystems and wildlife populations. The results inform decision-makers about whether any mitigation measures, such as species protection plans or habitat restoration initiatives, are necessary to reduce the project's impact [8].

4.3.2 Bird and Bat Mortality in Wind Energy

One of the primary concerns associated with wind energy development is its potential impact on bird and bat populations. Studies have shown that wind turbines, particularly those located along migratory routes or in areas with high bird or bat activity, can lead to fatalities due to collisions with the turbine blades. These fatalities are of particular concern for species already at risk of extinction, such as raptors and migratory birds, as well as bats that play vital roles in ecosystems by controlling insect populations.

Studies on Mortality Research has demonstrated that bird and bat fatalities tend to be highest in areas with significant concentrations of migratory species or near habitat-rich areas such as wetlands or forests. Additionally, certain species of bats are more vulnerable to turbine collisions, especially during the warmer months when they are more active. Mortality rates can vary depending on the size of the wind farm, turbine placement, and local environmental conditions [9].

Mitigation Strategies To reduce bird and bat mortality, several mitigation strategies have been developed and are implemented throughout the planning, construction, and operational phases of wind energy projects. One of the most effective mitigation methods is turbine curtailment, which involves temporarily shutting down turbines during periods of high bird or bat activity, such as during peak migration seasons or at dusk and dawn when bats are most active. Other mitigation measures include turbine design modifications, such as slowing the rotation of blades in certain conditions or altering turbine placement to avoid known migration corridors.

In addition, habitat restoration efforts near wind farms can help create alternative environments for wildlife, thereby reducing the risk of collisions. Monitoring bird and bat populations before, during, and after construction provides data to assess the effectiveness of mitigation strategies and informs adaptive management practices [10].

4.3.3 Habitat Fragmentation and Restoration

Renewable energy projects, like any large infrastructure developments, can lead to habitat fragmentation, which occurs when a large, continuous habitat is broken into smaller, isolated patches due to construction activities such as road-building, power line installation, and the development of wind and solar farms. Fragmentation can disrupt wildlife migration patterns, isolate populations, and reduce genetic diversity. It also increases edge effects, where the boundaries of habitat patches are subject to environmental conditions that are different from the interior, such as increased exposure to predators, human disturbance, or invasive species.

Assessment of Fragmentation Environmental assessments focus on evaluating how a project's infrastructure will affect existing wildlife corridors, migration routes, and access to key resources. The goal is to determine how the project might fragment habitats and whether it could create barriers to species movement. Advanced modeling techniques, such as Geographic Information Systems (GIS), can be used to identify critical wildlife corridors and predict how they might be impacted by proposed development.

Mitigation and Restoration Strategies To mitigate the effects of habitat fragmentation, developers can employ strategies such as wildlife corridors, which allow animals to safely travel between habitat patches. These corridors can be restored or constructed alongside renewable energy projects to maintain connectivity. Additionally, minimizing habitat disturbance during construction, and using pre-existing infrastructure (such as roads and transmission lines), helps to reduce the project's overall environmental footprint.

After construction, habitat restoration efforts are crucial for re-establishing ecological functions. These efforts might include replanting native vegetation, reintroducing species, or restoring hydrological conditions that were altered by the development [11].

4.3.4 Marine Wildlife in Offshore Renewable Projects

Offshore renewable energy projects, particularly offshore wind farms and tidal energy installations, are becoming increasingly important in the global transition to renewable energy. However, these projects present unique challenges for marine wildlife, including fish, marine mammals, and seabirds. Offshore wind farms, for example, can impact marine ecosystems in several ways, including noise pollution from construction, changes to underwater habitats, and the potential for collision-related mortality.

Considerations for Marine Wildlife Marine species, including commercially important fish species, cetaceans (whales and dolphins), and migratory seabirds, may be affected by offshore renewable energy projects. During the construction phase, underwater noise generated by pile-driving and other activities can disturb marine mammals, potentially leading to displacement from critical habitats or interference with communication and navigation. The physical presence of turbines and infrastructure can alter marine habitats, including benthic ecosystems (seafloor communities), and affect species that rely on these habitats for food or breeding.

Mitigation Strategies To minimize these impacts, developers use several strategies. During construction, **noise mitigation measures** such as bubble curtains or soft-start techniques (gradual increase in noise levels) can help reduce noise disturbance to marine mammals. **Seasonal restrictions** on construction activities can also be implemented to avoid critical breeding and migration periods for marine species.

In terms of habitat impacts, **environmentally sensitive siting** is a key strategy. Offshore wind farms are often located in areas where the potential to affect critical habitats is minimal. Developers may also design turbine structures to incorporate marine biodiversity by allowing natural reefs to form around them, enhancing local fish populations and promoting ecological resilience.

Monitoring and adaptive management are also key components of offshore projects. Regular surveys of marine life during and after construction help assess the success of mitigation measures and inform future practices [12].

4.4 Water Resource Assessment

Water resource assessments are crucial for understanding the potential impacts of renewable energy projects on local water bodies. Renewable energy systems, especially solar, hydroelectric, and wind projects, can interact with water resources in various ways. Evaluating the influence of these projects on surface water, groundwater, and aquatic ecosystems helps ensure that the benefits of renewable energy do not come at the cost of long-term environmental degradation.

4.4.1 Hydrological Studies

Hydrological studies are conducted to evaluate the impact of renewable energy projects on local water resources, including groundwater and surface water systems.

These studies focus on how the construction and operation of energy facilities could alter the natural flow of water, affect water quality, or change the availability of water for other uses such as agriculture, drinking, or recreational purposes.

Assessment Methodology Hydrological studies typically involve the collection of data on precipitation, runoff patterns, groundwater levels, and seasonal variations in water availability. Advanced hydrological modeling can predict how energy development might influence local hydrology. For example, in the case of wind farms or solar plants, the construction of roads, transmission lines, and other infrastructure can affect natural drainage patterns, potentially leading to erosion, sedimentation, or altered groundwater recharge rates.

For projects like hydroelectric plants, hydrological studies are crucial in assessing how the damming of rivers or alteration of water flow might impact the surrounding ecosystem. Hydropower projects can cause changes to river flow, water temperature, and sediment transport, which can, in turn, affect aquatic habitats and water quality. The findings from these studies inform the design of mitigation measures, such as the inclusion of fish ladders, altered dam release schedules, or adjustments to the layout of renewable energy infrastructure [13].

4.4.2 Water Usage in Solar Energy Projects

Water use is a significant consideration in certain types of solar energy projects, particularly **solar thermal plants**, which rely on large amounts of water for cooling and cleaning purposes. Solar thermal technologies, such as concentrated solar power (CSP), utilize mirrors or lenses to focus sunlight onto a receiver, where it generates heat that is used to produce electricity. This process often requires cooling systems that consume water, which can strain local water resources, especially in arid regions where CSP plants are commonly sited.

Water Needs for Cooling and Cleaning Cooling systems used in CSP plants typically rely on water to dissipate heat generated during the energy production process. Wet cooling systems, in particular, use large amounts of water to condense steam back into water, leading to a high rate of water consumption. Additionally, solar panels, whether used in photovoltaic (PV) systems or CSP plants, require regular cleaning to maintain efficiency. In areas with frequent dust accumulation or dry climates, cleaning can significantly increase water demand.

Environmental Implications The environmental implications of water use in solar energy projects depend on the local context, including water availability and competing demands. In water-scarce regions, the water required for cooling and cleaning could pose a challenge, especially if the resource is already under stress from agricultural, industrial, or domestic uses. To minimize this impact, some solar thermal plants are opting for **dry cooling technologies**, which require less water than traditional wet cooling systems. Others may use **alternative water sources**

such as brine or recycled water, reducing the burden on local freshwater supplies [14].

In addition to direct water use, the location of large-scale solar energy projects in regions with limited water availability must be carefully considered during the planning stages to avoid exacerbating water scarcity and ensure sustainable operation.

4.4.3 Protecting Aquatic Ecosystems in Hydroelectric Projects

Hydroelectric power, one of the oldest forms of renewable energy, harnesses the energy of flowing water to generate electricity. However, the construction of dams and reservoirs can have significant impacts on surrounding aquatic ecosystems, particularly rivers, lakes, and wetlands. These projects can alter water flow, water quality, sediment transport, and habitat availability, which may negatively affect fish populations, aquatic plants, and other species that rely on these habitats.

Measures to Safeguard Aquatic Ecosystems To protect aquatic ecosystems in hydroelectric projects, several mitigation measures can be employed:

1. **Fish Passage and Migration:** The construction of fish ladders or fish bypass systems ensures that migratory species, such as salmon, can continue their journey upstream or downstream despite the presence of dams. These systems are designed to allow fish to bypass barriers created by turbines and dams, promoting the survival of key species that depend on river systems for breeding and feeding.
2. **Flow Regulation:** Maintaining **environmentally responsible flow regimes** is essential to protect aquatic ecosystems. By releasing water in a way that mimics natural river flows, projects can reduce the adverse effects on fish spawning, aquatic plant life, and river sediment transport. This approach, called "environmental flow management," helps balance energy generation with the ecological health of the river system.
3. **Water Quality Monitoring:** Regular monitoring of water quality before, during, and after the construction of a hydroelectric plant is critical to detect and mitigate any adverse changes. Key parameters such as water temperature, dissolved oxygen, and pH must be closely monitored to ensure that the quality of water remains suitable for aquatic life. Elevated temperatures or low oxygen levels due to altered water flow can lead to the depletion of aquatic species and disrupt entire ecosystems.
4. **Sediment Management:** The disruption of sediment transport due to dam construction can impact the health of downstream aquatic ecosystems. Sediment is essential for maintaining the natural structure of riverbeds, providing habitats for aquatic organisms and contributing to soil fertility in surrounding areas. Strategies to mitigate sedimentation include the use of sediment bypass systems,

which allow sediment to pass downstream without being trapped in reservoirs, and regular dredging of reservoirs to manage sediment accumulation [15].

Through a combination of these measures, hydroelectric projects can be developed in a way that minimizes their environmental footprint while still generating renewable energy. However, the effectiveness of these strategies depends on careful planning, continuous monitoring, and adaptive management throughout the lifecycle of the project.

4.5 Noise and Light Pollution Assessment

As renewable energy projects grow in scale and geographic spread, their environmental impacts extend beyond air, water, and biodiversity. Noise and light pollution are significant considerations that can affect local communities, wildlife, and ecosystems. In renewable energy projects, careful planning and mitigation strategies are required to minimize these impacts.

4.5.1 Noise Impact of Wind Turbines

Wind turbines are a major source of renewable energy, but they are also associated with the generation of noise, which can have various consequences for local communities and wildlife. The noise produced by wind turbines is primarily a result of the movement of the blades through the air and the mechanical operations within the turbine itself.

Assessment of Noise Levels Noise assessments for wind energy projects are critical to understanding the potential impacts on nearby residents and wildlife. Wind turbine noise is generally classified into two types:

1. **Aerodynamic noise**, caused by the blades moving through the air.
2. **Mechanical noise**, originating from the turbine's mechanical components such as the gearbox and generator.

To assess the noise impact, studies measure the sound pressure levels at various distances from the turbine site. These measurements are taken during different wind conditions, as wind speed and direction can significantly influence the perception of turbine noise. Sound levels are typically measured in decibels (dB) and are compared with local noise regulations and thresholds to evaluate the extent of disturbance.

Impact on Local Communities For nearby communities, noise from wind turbines can be disruptive, particularly in rural or quiet areas. Wind turbine noise can cause annoyance, sleep disturbance, and other health impacts such as increased stress or anxiety. Noise assessments aim to predict these impacts and identify potential mitigation measures.

Impact on Wildlife Wildlife, especially nocturnal species, can also be impacted by turbine noise. Studies have shown that some bird and bat species may be displaced from habitats due to noise disturbances. The continuous low-frequency sounds generated by turbines can alter animal behavior, particularly for species that rely on sound for communication, navigation, and hunting. For example, bats are known to be sensitive to sound frequencies, and excessive noise could interfere with their echolocation abilities/ [16].

Mitigation Strategies Several strategies can be used to mitigate noise impact. These include:

- **Strategic siting** of turbines to minimize noise exposure to nearby communities and sensitive habitats.
- **Turbine design improvements**, such as the use of quieter blade designs and advanced sound-damping technologies to reduce mechanical noise.
- **Operational strategies**, such as adjusting the speed of turbine blades during periods of high noise sensitivity or implementing "curtailment" measures to temporarily shut down turbines during sensitive times (e.g., nighttime for local communities or during bat migration periods).

4.5.2 Light Pollution from Solar Farms

While solar energy is a clean and renewable resource, large-scale solar farms can contribute to light pollution, which affects nocturnal ecosystems and human health. Light pollution occurs when artificial light spills into areas that would otherwise be dark, disrupting the natural day-night cycle.

Evaluating the Impact of Solar Installations Solar farms, particularly large installations, can emit light pollution in two ways:

1. **Reflection of sunlight off panels** during early morning or late afternoon hours.
2. **Nighttime security lighting** used for the protection of the site during non-operational hours.

The light emitted from solar installations may not be as intense as artificial lights used in urban environments, but the cumulative impact of numerous solar projects in a region can contribute to a significant increase in ambient light levels. This light intrusion can disrupt nocturnal ecosystems, particularly for species sensitive to light, such as moths, bats, and certain bird species. Nocturnal animals rely on darkness for various critical behaviors, such as hunting, mating, and navigating. Excessive light can disorient these animals, leading to disruptions in their behavior and even their survival.

Impact on Nocturnal Ecosystems In addition to wildlife, light pollution can affect the human population, particularly in rural or natural areas, by diminishing the quality of the night sky and contributing to "skyglow," which impacts astronomical observation. For nocturnal animals, the presence of light can disrupt biological rhythms such as circadian cycles, potentially interfering with migration patterns or reproductive timing. For example, sea turtles, which rely on natural light cues for hatchling orientation, can be disoriented by artificial light sources, leading hatchlings away from the ocean and increasing mortality rates [17].

Mitigation Strategies Mitigating the light pollution caused by solar farms involves several approaches:

- **Orientation and Siting:** Ensuring that solar panels are oriented in a way that minimizes the reflection of sunlight into surrounding areas.
- **Lighting Controls:** Installing low-intensity, motion-sensor lighting that activates only when necessary, and ensuring that it is shielded to prevent light from spilling into the surrounding environment.
- **Site Selection:** Locating solar farms away from sensitive habitats and areas with high biodiversity to reduce the potential impacts on wildlife. Additionally, choosing sites with minimal human habitation can reduce the social impacts of light pollution.

By considering the potential for light pollution and adopting mitigation strategies, solar farms can reduce their environmental footprint while still contributing to the renewable energy transition.

4.6 Air Quality and Emissions

Renewable energy projects offer substantial environmental benefits, particularly in terms of air quality and emissions reduction. However, like any large infrastructure project, the construction and operational phases of renewable energy projects can have varying degrees of impact on air quality. These impacts need to be carefully assessed and mitigated to maximize the environmental benefits of renewable energy deployment.

4.6.1 Construction-Phase Emissions

The construction phase of renewable energy projects can result in temporary air quality degradation, primarily due to emissions from construction equipment, vehicle traffic, and other activities associated with land preparation and site development. These emissions typically include particulate matter (PM), nitrogen oxides (NOx), sulfur dioxide (SO2), and carbon monoxide (CO), which can affect local air quality and contribute to short-term environmental and health concerns.

Sources of Emissions

- **Construction Equipment:** Heavy machinery such as bulldozers, excavators, and cranes is used to prepare the site for renewable energy installations. These machines often run on diesel fuel and can emit significant amounts of particulate matter and NOx, contributing to air pollution.
- **Vehicle Traffic:** The movement of materials and personnel to and from the site increases vehicle emissions. Trucks and delivery vehicles may contribute to the release of NOx and particulate matter, especially in areas with high traffic density or on unpaved roads.
- **Site Preparation Activities:** Dust from land grading, excavation, and other site preparation activities can lead to the release of particulate matter, which negatively impacts air quality, particularly in areas with arid conditions or high wind speeds.

Mitigating Construction-Phase Emissions To minimize the air quality impacts during the construction phase, several strategies can be employed:

- **Dust Control Measures:** Implementing dust control measures such as water spraying, dust suppressants, and covering stockpiled materials can help reduce particulate emissions from site preparation.
- **Low-Emission Equipment:** Using low-emission construction equipment, such as those with diesel particulate filters or electric-powered machines, can reduce the amount of NOx and particulate matter released.
- **Traffic Management:** Limiting the number of vehicles on-site, using routes with less traffic congestion, and encouraging the use of fuel-efficient vehicles can help mitigate vehicle-related emissions.
- **Construction Scheduling:** Scheduling construction activities during times of lower air quality or wind speeds can minimize the dispersion of pollutants into the surrounding environment.

By implementing these mitigation strategies, the negative air quality impacts of construction activities can be reduced, helping to ensure that the environmental benefits of renewable energy are realized without significant short-term harm to air quality.

4.6.2 Long-Term Benefits for Air Quality

One of the primary advantages of renewable energy projects is their long-term positive impact on air quality. Unlike fossil fuel-based energy generation, which emits large quantities of air pollutants, renewable energy technologies such as wind, solar, and hydropower have little to no direct emissions once operational.

The transition to renewable energy can lead to substantial reductions in harmful pollutants such as carbon dioxide (CO_2), sulfur dioxide (SO_2), and nitrogen oxides (NO_x), all of which are major contributors to air pollution, smog, and acid rain.

Reduction in Greenhouse Gas Emissions The most significant long-term benefit of renewable energy for air quality is the reduction in greenhouse gas (GHG) emissions. Fossil fuel-based power plants are major sources of CO_2, the leading greenhouse gas responsible for climate change. By replacing coal, oil, and natural gas with wind, solar, and hydropower, renewable energy projects help reduce the overall carbon footprint of energy production.

Reduction in Other Air Pollutants Renewable energy projects also help reduce other harmful air pollutants that contribute to smog, respiratory issues, and environmental degradation. For instance, the combustion of fossil fuels in power plants releases significant quantities of SO2 and NOx, both of which contribute to acid rain and the formation of ground-level ozone, a harmful air pollutant. By displacing fossil fuel-based energy sources, renewable energy projects reduce the emissions of these pollutants, leading to cleaner air.

Health Benefits The reduction in air pollution has direct health benefits for local populations. Air pollution is linked to a variety of respiratory and cardiovascular diseases, including asthma, bronchitis, and heart disease. By improving air quality, renewable energy projects can reduce the incidence of these health conditions, leading to a healthier population. According to a study by the Clean Energy Trust [18], the widespread deployment of renewable energy could prevent thousands of premature deaths and reduce medical costs associated with air pollution.

Regional Air Quality Improvements Renewable energy deployment also helps improve air quality on a regional scale. Areas that rely heavily on fossil fuels, such as coal-burning power plants, often experience poor air quality and frequent smog. Shifting to renewable energy sources can help alleviate these issues and provide cleaner air for both urban and rural populations. For example, large-scale wind and solar farms in regions that currently depend on coal for power can lead to significant reductions in regional air pollution, improving public health and environmental quality.

Global Air Quality Benefits On a global scale, the transition to renewable energy is a key strategy for mitigating climate change and improving overall air quality. The widespread adoption of renewable energy can help reduce the global reliance on fossil fuels, leading to a significant decrease in CO_2 and other air pollutants. This global reduction in emissions can help limit the severity of climate change and reduce the environmental and health impacts associated with poor air quality.

In conclusion, while renewable energy projects may cause some temporary air quality degradation during the construction phase, their long-term benefits far outweigh these short-term impacts. By reducing emissions of harmful pollutants, renewable energy technologies play a critical role in improving air quality, combating climate change, and promoting public health.

4.7 Social Impact and Community Engagement

Renewable energy projects not only transform energy systems but also bring significant changes to the social and cultural dynamics of the communities in which they are developed. Understanding these impacts and fostering meaningful community engagement are essential for the successful deployment of projects, ensuring that they deliver equitable benefits while respecting local cultures and ecosystems.

4.7.1 Understanding Social and Cultural Impacts

Renewable energy projects can have diverse social and cultural implications, influencing the lives of local residents and the integrity of their heritage.

4.7.1.1 Impact on Local Communities and Cultures

Renewable energy developments can alter the social fabric of local communities. Large-scale projects may require land acquisition, leading to changes in land use, displacement of residents, or restricted access to community resources [19]. These changes can disrupt traditional lifestyles and economic practices, especially in rural or indigenous areas where people rely on the land for their livelihoods.

Cultural shifts may also occur as infrastructure and workforce migration introduce new dynamics to the community. For example, construction crews and operational staff from outside the region may influence local social structures, both positively (through increased diversity) and negatively (through cultural friction).

4.7.1.2 Cultural Resources and Heritage

Energy projects can impact archaeological sites, historical landmarks, and other cultural heritage areas. Ground-disturbing activities for wind turbine installation or solar farm construction may inadvertently damage ancient artifacts or sacred sites [20].

To address this, developers must conduct detailed cultural resource assessments and consult with local stakeholders, including indigenous groups, to identify and protect these valuable assets. Mitigation strategies may involve altering project designs, creating buffer zones, or funding the preservation of cultural resources.

4.7.2 *Economic and Employment Impacts*

Renewable energy projects bring both economic opportunities and challenges to local communities.

4.7.2.1 Job Creation and Economic Benefits

The renewable energy sector is a significant driver of job creation. From construction and installation to maintenance and operations, projects generate employment across a range of skill levels. A study by the International Renewable Energy Agency[21] estimates that global employment in the renewable energy sector could exceed 42 million by 2050.

In addition to jobs, renewable energy projects stimulate local economies through increased demand for goods and services, including housing, food, and transportation for workers. Communities often benefit from infrastructure improvements, such as better roads or access to clean energy, which enhance overall quality of life.

4.7.2.2 Potential Adverse Effects on Existing Local Industries

While renewable energy can bring economic benefits, it may also compete with or disrupt existing industries. For example, agricultural lands used for solar farms may reduce food production, affecting local farmers. Similarly, the visual impact of wind turbines can deter tourism in areas known for scenic beauty [22]. Mitigating these effects involves careful project siting, stakeholder engagement, and the exploration of synergies between renewable energy and existing industries (e.g., agrivoltaics[1]).

4.7.3 *Importance of Engaging Local Communities and Social License to Operate*

Successful renewable energy projects depend on building trust and goodwill with local communities. This involves meaningful engagement to ensure that projects align with community needs and values.

4.7.3.1 Community Consultations and Feedback Mechanisms

Engaging local communities early and throughout the project lifecycle is critical. Community consultations allow developers to share project goals, address concerns,

[1] Agrivoltaics is the dual use of land for both solar energy generation and agricultural production, maximizing land efficiency and sustainability.

and incorporate local knowledge into decision-making. Transparent communication about potential impacts and benefits fosters trust and encourages community buy-in [23].

Feedback mechanisms, such as public forums or grievance redressal systems, empower communities to voice their opinions and influence project outcomes. This collaborative approach helps resolve conflicts, reduces opposition, and enhances project sustainability.

4.7.3.2 Social License to Operate

The concept of a "social license to operate" (SLO) refers to the informal approval and ongoing support of a project by local stakeholders. Unlike regulatory approvals, an SLO is granted by communities and must be earned through genuine engagement, mutual respect, and demonstrated benefits [24].

Achieving an SLO involves delivering tangible economic and social benefits, minimizing negative impacts, and upholding ethical practices. For instance, developers who invest in community development initiatives, such as funding schools or healthcare facilities, are more likely to gain community trust. An SLO is not static; it must be maintained through continuous dialogue and responsiveness to evolving community needs.

4.7.4 Case Study: Community Engagement in the United States

4.7.4.1 Block Island Wind Farm, Rhode Island

The Block Island Wind Farm (Fig. 4.1), the first offshore wind project in the U.S., owned and operated by Ørsted US Offshore Wind, highlights the importance of community engagement and addressing social impacts. The developers worked closely with local stakeholders, including fishermen, residents, and environmental groups, to address concerns about the project's impact on marine life and coastal aesthetics.

One significant issue was the potential disruption to fishing activities. To mitigate this, the developers collaborated with the local fishing community to modify project plans and provide compensation for any losses. Transparent communication and public meetings ensured that stakeholders were informed and involved in decision-making processes.

The project delivered substantial benefits, including reducing reliance on diesel generators for electricity on Block Island, cutting greenhouse gas emissions, and lowering electricity costs for residents. This engagement-driven approach not only earned community support but also set a precedent for future offshore wind projects in the U.S. [26].

In conclusion, understanding and addressing the social and cultural impacts of renewable energy projects is crucial to equitable and sustainable development. By prioritizing community engagement, mitigating adverse effects, and delivering

Fig. 4.1 Block Island Wind Farm [25]

benefits, developers can secure the social license to operate. U.S.-based examples, such as the Block Island Wind Farm, demonstrate that a collaborative approach fosters trust, enhances project acceptance, and contributes to broader goals of social equity and environmental stewardship.

4.8 Mitigation of Environmental Risks

Renewable energy projects inherently aim to reduce global reliance on fossil fuels and minimize carbon footprints. However, they also come with localized environmental challenges, necessitating comprehensive mitigation measures. These efforts ensure that projects are not only sustainable but also harmonious with the ecosystems and communities they inhabit.

4.8.1 Categorizing Environmental Risks and Risk Prioritization

Categorizing risks involves identifying environmental hazards such as habitat disruption, species endangerment, water usage, noise, and air quality degradation. Tools such as Environmental Risk Assessment (ERA) frameworks enable developers to systematically identify and evaluate risks, providing a structured approach for mitigation planning [27].

Key Steps in Risk Prioritization

1. **Likelihood Assessment:** Evaluating how probable a specific impact is, given the project's location, technology, and scale.
2. **Consequence Evaluation:** Measuring the severity of potential impacts on ecosystems, species, and local communities.
3. **Mapping High-Risk Areas:** Utilizing Geographic Information Systems (GIS) to overlay environmental sensitivity with project infrastructure to avoid ecologically critical zones [28].
4. **Stakeholder Consultation:** Engaging experts, regulatory bodies, and local communities to identify risks that might not be immediately apparent.

This systematic approach allows developers to focus resources on addressing the most significant risks while ensuring regulatory compliance.

4.8.2 Mitigation Strategies and Best Practices

4.8.2.1 Engineering and Technological Solutions

1. **Wind Turbine Design Innovations:**
   Advances in turbine technology, such as radar-enabled curtailment systems, reduce avian collisions by slowing or stopping turbines when birds are detected nearby [29].
2. **Solar Panel Optimization:**
   The use of anti-glare coatings minimizes light pollution, which can confuse nocturnal species. Additionally, bifacial panels maximize land efficiency, reducing the need for habitat encroachment [30].
3. **Battery Recycling Technologies:**
   Ensuring the sustainability of energy storage involves innovations in lithium-ion battery recycling to prevent toxic leaching into the environment [31].

4.8.2.2 Operational and Management Practices

1. **Habitat Restoration Projects:**
   During construction, efforts such as planting native vegetation and creating wildlife corridors mitigate habitat fragmentation. For instance, the Solar Star Project implemented such measures to protect the desert tortoise [32].
2. **Adaptive Management Plans:**
   Using real-time monitoring systems, operators adjust activities based on observed impacts. For example, marine mammal presence near offshore wind

Fig. 4.2 Altamont Pass Wind Turbines at sunset [35]

farms may temporarily halt pile-driving activities to reduce underwater noise pollution [33].

3. **End-of-Life Management:**
   Creating decommissioning plans ensures that turbines, solar panels, and other infrastructure are dismantled responsibly, with components recycled or repurposed.

4.8.3 Case Studies

4.8.3.1 Altamont Pass Wind Resource Area, California

The Altamont Pass (Fig. 4.2), one of the oldest wind farms in the U.S., developed by Altamont Winds LLC, was initially criticized for high bird mortality rates specially the Hoary bat (Fig. 4.3). Mitigation measures included replacing outdated turbines with fewer, larger, and more efficient ones, strategically sited to avoid raptor migration routes. These actions resulted in a substantial reduction in avian fatalities, setting a precedent for risk mitigation in wind energy projects [34].

4.8.3.2 Desert Sunlight Solar Farm, California

The Desert Sunlight Solar Farm (Fig. 4.4), located in Riverside County, California, is one of the largest photovoltaic solar farms in the United States, with a capacity of 550 MW. Developed by NextEra Energy Resources and GE Financial Services, the project was constructed on public lands managed by the Bureau of Land Management (BLM).

Fig. 4.3 Hoary bat [35]

Fig. 4.4 Desert Sunlight Solar Farm [35]

While the solar farm provides substantial clean energy benefits, it faced environmental concerns due to its proximity to sensitive desert ecosystems, including habitats of the Mojave fringe-toed lizard and the desert tortoise (Fig. 4.5). To mitigate these impacts, developers implemented several measures: they translocated desert tortoises from the construction area, restored disturbed habitats, and carefully planned the layout to avoid ecologically critical zones.

In addition, the project underwent rigorous environmental review in collaboration with the U.S. Fish and Wildlife Service and local tribal representatives. These mitigation efforts exemplify how utility-scale solar projects can be aligned with conservation priorities when developed responsibly [36].

Fig. 4.5 Desert Tortoise [37]

4.9 Sustainable Development Goals (SDGs) and Renewable Energy

Renewable energy development is pivotal to achieving the United Nations' Sustainable Development Goals (SDGs), particularly those addressing energy, climate, and infrastructure. By aligning renewable energy projects with these global goals, developers contribute to sustainable development while ensuring long-term benefits for society and the planet.

4.9.1 Contributions to Specific SDGs

4.9.1.1 SDG 7: Affordable and Clean Energy

Renewable energy projects directly support SDG 7 by increasing access to affordable, reliable, and modern energy systems. Solar, wind, and other renewables enable energy access in remote areas, reducing dependency on traditional fossil fuels. For example, community solar programs in rural parts of the U.S. provide low-income households with affordable electricity, promoting energy equity [31].

SDG 13: Climate Action
Transitioning to renewable energy is central to addressing climate change by reducing greenhouse gas (GHG) emissions. Projects like the Ivanpah Solar Electric Generating System exemplify the role of renewables in offsetting emissions, displacing millions of metric tons of CO_2 annually (CEC, 2021). Additionally,

renewable energy serves as a foundation for net-zero carbon strategies and global climate commitments.

SDG 9: Industry, Innovation, and Infrastructure
Renewable energy projects contribute to building resilient infrastructure and fostering innovation. The development of floating solar farms, offshore wind platforms, and advanced battery storage exemplifies how renewable energy drives technological advancement and enhances grid reliability. Furthermore, these innovations often lead to job creation and economic development, benefiting both industry and local communities.

4.9.1.2 Integration of SDGs in Project Planning and Implementation

SDG-Aligned Project Goals
Setting SDG-aligned objectives during the planning phase ensures that projects address critical sustainability challenges. For instance, incorporating energy equity (SDG 7) into design processes can prioritize underserved areas for renewable installations.

Stakeholder Engagement and SDGs
Engaging stakeholders, including governments, NGOs, and local communities, ensures projects address SDG priorities relevant to the region. Collaboration fosters trust and aligns project outcomes with broader community and environmental goals [38].

Monitoring and Reporting on SDG Impact
Using frameworks like the Global Reporting Initiative (GRI) or SDG Compass, developers can measure and report their contributions to SDGs. Metrics such as GHG reductions, number of households electrified, and local economic benefits showcase progress toward these goals.

4.9.1.3 Challenges and Opportunities in Aligning Renewable Energy with SDGs

Challenges

1. **Policy and Regulatory Barriers:**
   Complex permitting processes and inconsistent policies can hinder project alignment with SDGs, particularly in achieving affordable energy access (SDG 7).
2. **Financing and Investment Gaps:**
   Limited funding for renewable projects in underserved regions may restrict equitable progress toward clean energy access.his challenge appears to be growing amid the current policy environment and the stagnation of renewable projects in grid queues, such as in Pennsylvania, where prolonged delays further hamper project development and investor confidence.

3. **Conflict Between SDGs:**
   Some SDGs, like economic growth (SDG 8), may inadvertently conflict with environmental conservation goals (SDG 15), requiring careful balancing in project design.

Opportunities

1. **Leveraging Innovation:**
   Emerging technologies like AI-driven energy optimization and green hydrogen production can advance multiple SDGs simultaneously.
2. **Scaling Community Engagement:**
   Inclusive consultation processes empower communities and ensure project benefits align with local priorities, creating long-term positive impacts.
3. **Collaborative Partnerships:**
   Public-private partnerships and global collaborations can drive large-scale renewable energy adoption while meeting SDG targets.

4.9.1.4 Case Study: Block Island Wind Farm, Rhode Island

The Block Island Wind Farm, as described in Sect. 4.7.4, exemplifies SDG integration:

- **SDG 7:** Replaced diesel generators on Block Island, providing clean and reliable energy.
- **SDG 13:** Reduced GHG emissions by eliminating reliance on fossil fuels.
- **SDG 9:** Pioneered offshore wind energy infrastructure in the U.S., spurring innovation and creating local jobs [39].

References

1. U.S. Environmental Protection Agency, Wetlands and the clean water act. https://www.epa.gov/cwa-404 (2019)
2. U.S. Army Corps of Engineers, Wetland delineation procedures. https://www.usace.army.mil (2022)
3. Ramsar Convention Secretariat, The ramsar convention on wetlands: international treaties for wetland conservation. https://www.ramsar.org (2020)
4. California Energy Commission, Solar energy and wetlands: best practices for development. https://www.energy.ca.gov (2018)
5. U.S. Department of Energy, Wind energy and the environment: addressing impacts and mitigation strategies. https://www.energy.gov/eere/wind/wind-energy-environment (2020)
6. International Hydropower Association, Hydropower and the environment: managing risks and opportunities. https://www.hydropower.org (2021)
7. U.S. Environmental Protection Agency, Wetland delineation and mapping. https://www.epa.gov/wetlands/wetland-delineation (2020)
8. J. Jones, Protected species and habitat surveys: guidelines for renewable energy projects. Environ. Sci. J. **34**(2), 119–130 (2020)

9. K.S. Smallwood, Impacts of wind energy on birds and bats: a review of mortality and mitigation strategies. Environ. Stud. J. **39**(1), 10–22 (2013)
10. E.B. Arnett, W.K. Brown, W.P. Erickson, J.K. Fiedler, B.L. Hamilton, T.H. Henry, Patterns of bat fatalities at wind energy facilities in North America. J. Wildl. Manag. **72**(5), 61–77 (2008)
11. M.L. Knight, J.E. McKechnie, R.S. Jones, Impacts of habitat fragmentation on wildlife: a case study in renewable energy development. Ecol. Restor. J. **15**(3), 45–56 (2019)
12. J.J. Roberts, D.S. Goehring, R.J. Bassett, Marine wildlife considerations for offshore renewable energy projects. Mar. Environ. Res. **25**(6), 273–285 (2020)
13. N.L. Poff, J.D. Olden, D.M. Merritt, D.M. Pepin, The ecological effects of dam removal and river channelization on biodiversity. BioScience **60**(10), 735–746 (2010)
14. E. Lantz, S. Tegen, J. Zuboy, Water use in renewable energy power plants. National Renewable Energy Laboratory (2016)
15. M. Dynesius, C. Nilsson, Fragmentation and flow regulation of river systems in the northern third of the world. Science **266**(5186), 753–762 (1994)
16. E.F. Baerwald, J.M. Edworthy, M. Holder, A large-scale mitigation experiment to reduce bat fatalities at wind energy facilities. J. Wildl. Manag. **72**(4), 1081–1091 (2008)
17. B.E. Witherington, R.E. Martin, Understanding, assessing, and resolving light-pollution problems on sea turtle nesting beaches. Florida Marine Research Institute Technical Report No (2000)
18. Clean Energy Trust, The health and economic benefits of renewable energy. Retrieved from www (2017)
19. K. Yenneti, R. Day, O. Golubchikov, Spatial justice and the land politics of renewables: dispossessing vulnerable communities through solar energy megaprojects. Geoforum **76**, 90–99 (2016)
20. S. King, P. Bridgewater, M. Smith, Cultural heritage and renewable energy landscapes. Landsc. Res. **43**(8), 1156–1171 (2018)
21. International Renewable Energy Agency (IRENA), World Energy Transitions Outlook: 1.5°C Pathway. Retrieved from www (2021)
22. M.C. Slattery, B.L. Johnson, J. Swofford, Public attitudes toward wind power in Texas: local communities in a renewable energy boom. Energy Policy **39**(5), 2320–2330 (2012)
23. N. Hall, P. Ashworth, P. Devine-Wright, Social acceptance of renewable energy projects: a conceptual framework. Energy Policy **61**, 268–277 (2013)
24. K. Moffat, A. Zhang, The paths to social license to operate: an integrative model explaining community acceptance of mining. Resourc. Policy **39**, 61–70 (2014)
25. Landscape Architecture Foundation, Block island wind farm - aerial image. https://www.landscapeperformance.org/case-study-briefs/block-island-wind-farm (2021). Accessed 02 May 2025
26. W. Musial, P. Beiter, S. Tegen, Block Island Wind Farm: America's first offshore wind project. National Renewable Energy Laboratory (2021)
27. U.S. Environmental Protection Agency, Environmental risk assessment. https://www.epa.gov/risk (2023). Accessed 02 May 2025
28. Esri, Gis in environmental risk management. https://www.esri.com (2023). Retrieved from Esri
29. T. Katzner et al., Mitigating avian impacts in wind energy. Conserv. Biol. **34**(1), 15–24 (2020)
30. R. Spencer et al., Floating solar panels: innovations and benefits. Environ. Sci. Lett. **15**(2), 231–250 (2022)
31. IRENA, Renewable power generation costs in 2019. International Renewable Energy Agency (2020)
32. U.S. Fish and Wildlife Service, Solar energy and wildlife habitat conservation. https://www.fws.gov (2023). Retrieved from USFWS
33. R. Wiser et al., Monitoring and adaptive management in renewable energy projects. Energy Policy **135**, 111045 (2021)
34. K.S. Smallwood, B. Karas, Avian mortality in the Altamont Pass wind resource area. J. Wildl. Manag. **73**(7), 100–115 (2009)

35. L.M. Krieger, Bay area wind turbines pose threat to 'George Clooney' of the bat world (2022). Accessed 02 May 2025
36. U.S. Bureau of Land Management, Desert sunlight solar farm (2022). Accessed 02 May 2025
37. W. Driscoll, A desert solar farm can actually improve desert tortoise habitat (2020). Accessed 03 May 2025
38. United Nations Development Programme, Integrating sdgs into energy projects. https://www.undp.org (2022). Retrieved from UNDP
39. U.S. Department of Energy, Offshore wind energy: Block island case study. https://www.energy.gov (2023). Retrieved from DOE

Chapter 5
Technological Innovations in Renewable Energy

5.1 Advances in Solar PV and CSP Technologies

The renewable energy sector has seen significant strides in solar photovoltaic (PV) and concentrated solar power (CSP) technologies. These innovations are pivotal for improving efficiency, reducing costs, and enhancing the scalability of solar energy systems. This section explores key advancements in these technologies, including bifacial and tandem solar cells, thin-film and perovskite solar cells, and CSP systems with thermal energy storage.

5.1.1 Bifacial Solar Cells

- **Efficiency Improvements in Bifacial Solar Panels:**
  Bifacial solar panels are designed to capture sunlight on both the front and rear sides of the panel, increasing energy production. By using reflective surfaces like white roofs or light-colored ground materials, the rear side of the panel can harvest additional light reflected from the environment (Fig. 5.1).

Recent advancements in bifacial technology have led to efficiency improvements, with some models now achieving energy yield increases of 10–30% compared to traditional monofacial panels. Enhanced materials, better rear-side encapsulation, and optimized panel designs have all contributed to these gains (Fig. 5.2).

5.1.2 Tandem Cell Technology for Better Energy Conversion

Tandem solar cells involve stacking multiple layers of solar cells made from different materials, each designed to capture a specific part of the solar spectrum. The most promising configuration combines silicon with other materials, such as

D. Gupta et al., *The Roadmap to Renewable Energy*,
https://doi.org/10.1007/978-3-032-09609-8_5

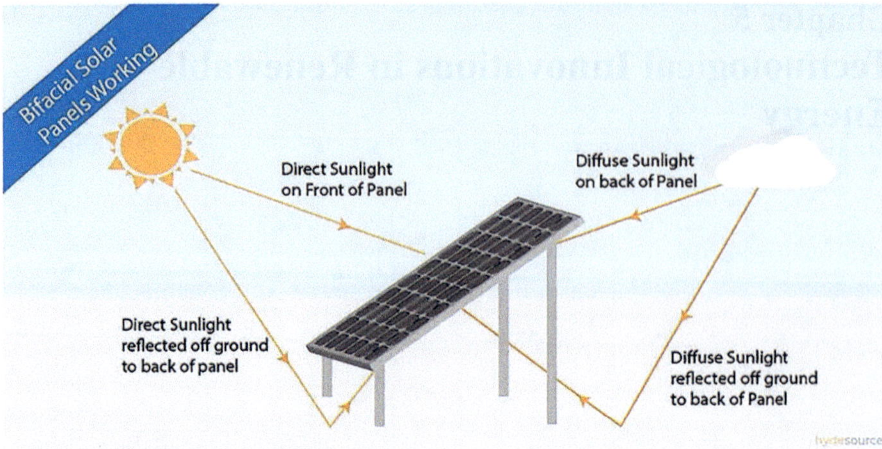

Fig. 5.1 Working of bifacial panel [1]

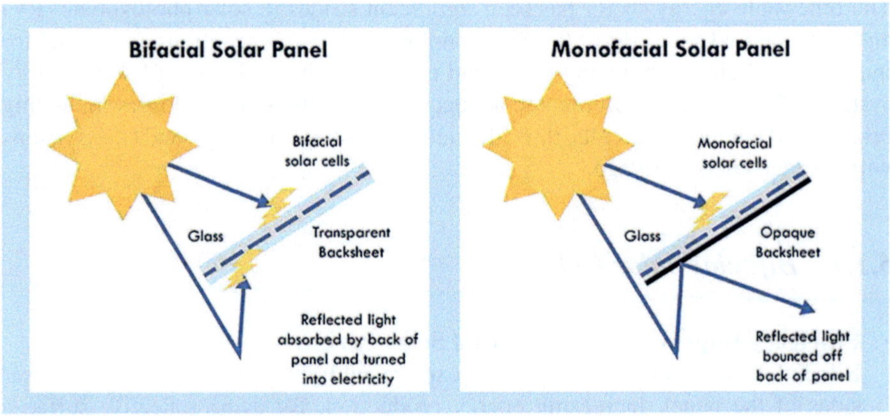

Fig. 5.2 Bifacial vs monofacial [2]

perovskite, to create a more efficient multi-junction solar cell (Fig. 5.3). These cells can theoretically exceed the efficiency limits of conventional silicon cells, with efficiencies reaching above 30%. Tandem solar cells are a breakthrough in improving solar energy conversion, as they enable the capture of a broader range of light wavelengths, resulting in higher power output.

These advancements pave the way for further innovations in photovoltaic technologies, particularly in thin-film and Perovskite solar cells, which offer additional advantages in terms of flexibility, weight, and cost.

- **Innovations in Thin-Film Solar Materials for Lightweight Applications:** Thin-film solar cells (Fig. 5.4) have been at the forefront of innovation for applications requiring lightweight and flexible materials. Unlike traditional silicon solar panels, thin-film cells are made by depositing one or more layers

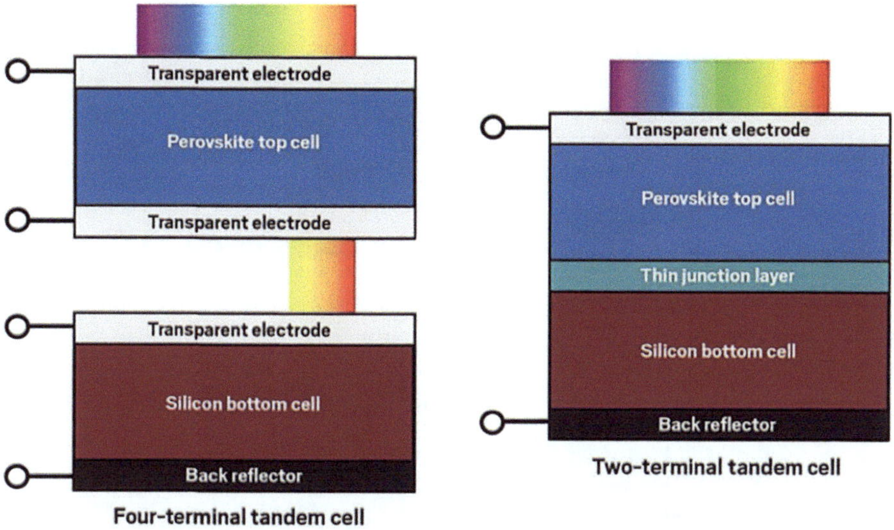

Fig. 5.3 Tandem cell stacking [3]

Fig. 5.4 Thin film solar cell [5]

of photovoltaic material onto a substrate. They offer several advantages, such as lower production costs, flexibility, and light weight, making them ideal for integration into a variety of surfaces, including building-integrated photovoltaics (BIPV) and portable applications. Recent advancements focus on enhancing the efficiency and longevity of thin-film materials, such as cadmium telluride (CdTe) and copper indium gallium selenide (CIGS).

Companies and labs (like Oxford PV, Heliatek, and NREL) are actively developing ultrathin tandem modules with >30% efficiency [4].

These can be rolled, folded, or applied like stickers to surfaces.

- **Emerging Perovskite Solar Cells and Their Potential for Higher Efficiency:** Perovskite solar cells are one of the most promising emerging technologies in the solar industry. These cells use a unique crystal structure that allows for highly efficient light absorption and conversion (Fig. 5.5). In recent years,

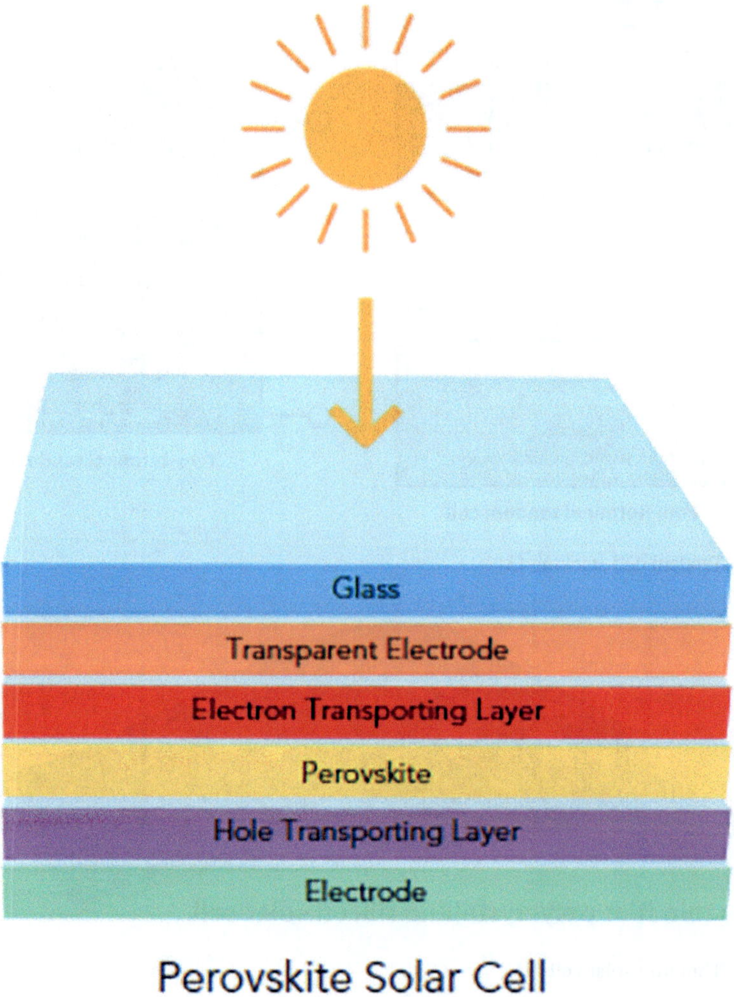

Fig. 5.5 Perovskite solar cells

perovskite solar cells have shown remarkable progress, with efficiencies soaring from around 3% in the early 2000s to over 25% in lab settings. Their potential lies in both their ability to be manufactured at low cost and their flexibility, enabling them to be used in various applications, including on windows and other unconventional surfaces. However, challenges remain in improving their stability and scaling up production while addressing environmental concerns associated with their lead-based components.

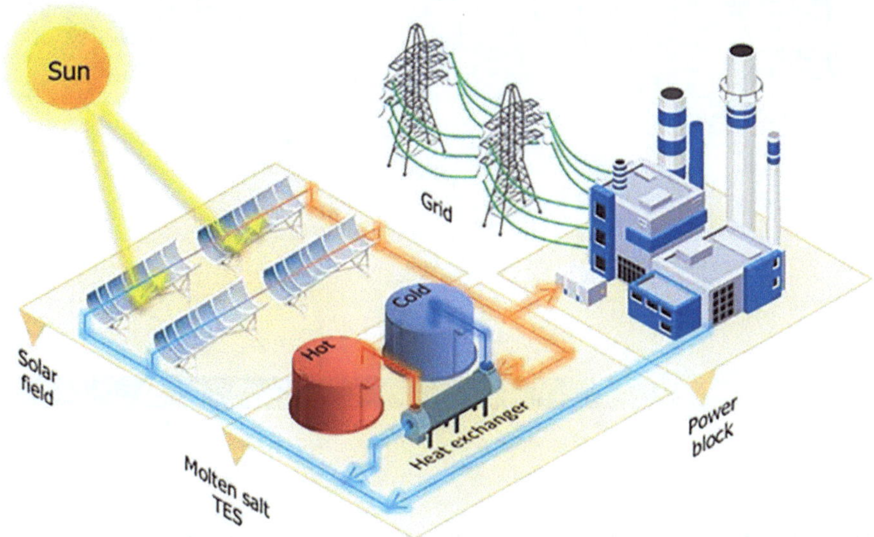

Fig. 5.6 Diagram of a parabolic trough collector (PTC) technology CSP plant with a TES system [6]

5.1.3 CSP with Thermal Energy Storage

5.1.3.1 Advances in Concentrated Solar Power (CSP) Combined with Molten Salt and Thermal Energy Storage Systems

Concentrated Solar Power (CSP) technologies use mirrors or lenses to focus sunlight onto a small area, generating intense heat that is converted into electricity. Recent innovations have focused on integrating CSP with advanced thermal energy storage (TES) systems, such as molten salt (Fig. 5.6). Molten salt storage allows CSP plants to store thermal energy for hours or even days, enabling them to produce electricity even when the sun is not shining. This enhances the reliability and dispatchability of CSP systems, addressing one of the key challenges of solar power: intermittency. By improving storage efficiency, increasing the scalability of molten salt systems, and reducing costs, CSP combined with TES is positioned to play a significant role in future renewable energy grids. These technological advancements in solar PV and CSP are transforming the landscape of renewable energy. As efficiency improves and costs decrease, solar technologies are becoming more competitive with conventional energy sources, driving the global shift towards cleaner, more sustainable energy solutions.

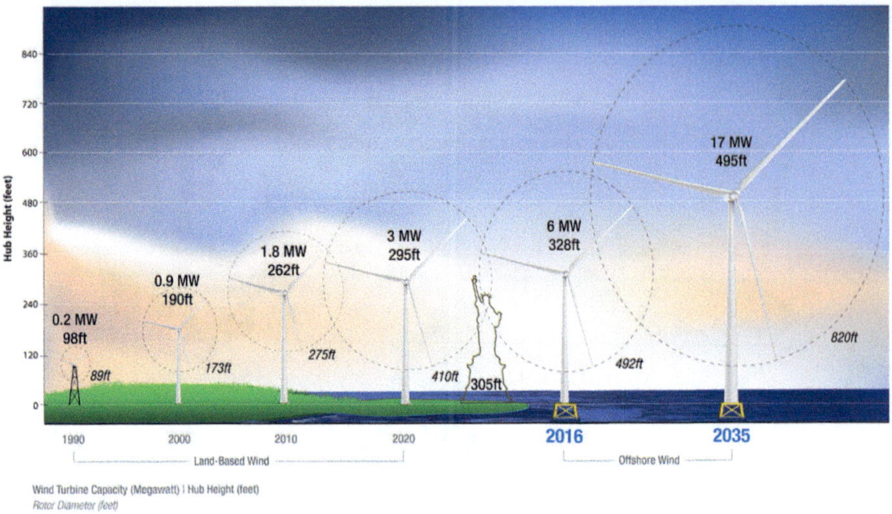

Fig. 5.7 Comparison of rotor blade sizes and their respective swept areas [8]

5.2 Wind Turbine Design Innovations

5.2.1 Larger Rotor Blades and Turbine Scaling

5.2.1.1 Benefits of Larger Rotor Blades in Increasing Power Output

Larger rotor blades have transformed the wind energy industry by allowing turbines to capture more wind energy, thereby significantly increasing their power output (Fig. 5.7). The amount of energy a wind turbine generates is proportional to the swept area of its rotor blades, which grows quadratically with blade length. For example, doubling the blade length results in a fourfold increase in the swept area, vastly improving energy capture [7].

These larger blades enhance energy production at lower wind speeds, making it feasible to deploy wind turbines in areas with less consistent wind resources. This capability has expanded the geographical viability of wind farms, reducing reliance on traditional high-wind locations like coastal areas or mountain ranges [9].

In addition to improved energy capture, larger blades reduce the overall cost per megawatt-hour of electricity generated. By increasing the capacity of individual turbines, fewer turbines are needed to achieve the same energy output. This translates to lower installation and maintenance costs, as well as reduced land use, making wind farms more cost-efficient and environmentally sustainable [10].

Fig. 5.8 UK's Largest Onshore Turbine Blades, transport to Muirhall Wind Farm, South Lanarkshire [12]

5.2.1.2 Engineering Challenges and Advancements in Scaling Turbine Sizes

The scaling up of rotor blades introduces several engineering and logistical challenges. First, larger blades are subjected to greater aerodynamic forces, leading to higher stress on the turbine's structural components. The increased weight and load on the hub, nacelle, and tower demand more robust materials and engineering solutions to maintain structural integrity and operational safety.

To overcome these challenges, manufacturers are developing advanced materials such as carbon-fiber-reinforced composites, which are both lightweight and exceptionally strong. These materials not only reduce blade weight but also improve resistance to fatigue and environmental conditions like high winds and extreme temperatures [11].

Another major obstacle is the transportation of these enormous blades, particularly for onshore wind farms. Figure 5.8 shows a turbine blade being transported to Muirhall Wind Farm.

Blades that exceed 70 m in length are difficult to navigate through roads and urban infrastructure. To address this, companies are adopting modular blade designs that can be assembled on-site, thereby circumventing transportation constraints [9].

Advances in aerodynamics also play a critical role in scaling turbine sizes. Computational fluid dynamics (CFD) simulations have led to optimized blade shapes that maximize energy capture while minimizing drag and turbulence. Features like serrated trailing edges and adaptive pitch control have further enhanced the efficiency and reliability of larger turbines [7].

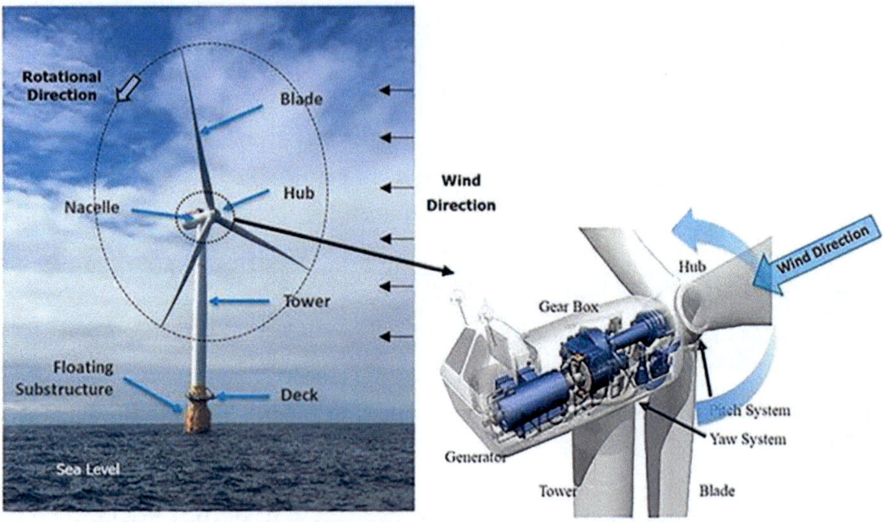

Fig. 5.9 A typical offshore wind turbine

5.2.1.3 Innovations in Offshore Turbines

While scaling turbines is beneficial for land-based wind farms, it is even more crucial for offshore installations, where wind speeds are typically higher and more consistent. The deployment of larger turbines offshore reduces the cost of installation and maintenance, as fewer turbines are needed to generate the same amount of energy. This has led to the development of turbines with rotor diameters exceeding 200 m, designed specifically for offshore applications [10]. A typical structure of an offshore turbine is shown in Fig. 5.9.

These innovations in turbine scaling are expected to play a vital role in achieving global renewable energy targets. By addressing the challenges associated with larger rotor blades, the wind energy sector can continue to expand its contribution to the global energy mix while driving down costs and improving sustainability.

5.2.2 Innovations in Floating Wind Platforms for Deep-Sea Installations

The development of floating offshore wind turbines has revolutionized the renewable energy landscape by allowing access to untapped wind resources in deep-sea locations. These platforms, unlike fixed-bottom turbines, are designed to operate efficiently in water depths greater than 60 m, where traditional installations are impractical due to geological and economic constraints [13].

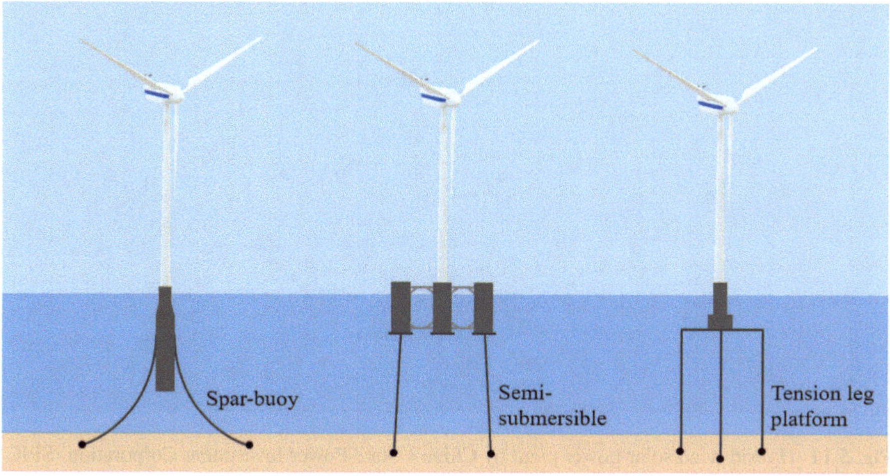

Fig. 5.10 Three major type of floating offshore wind turbines

1. **Platform Designs and Structural Advances**
   Floating platforms are categorized into spar-buoy, semi-submersible, and tension-leg platforms (Fig. 5.10) [13]. Each design is tailored to specific maritime conditions: spar-buoys rely on a ballast-stabilized keel for deep waters, semi-submersible platforms distribute weight over multiple pontoons to maintain buoyancy, and tension-leg platforms use taut mooring systems for added stability in dynamic wave environments. Recent advances in composite materials, such as carbon-fiber-reinforced polymers, have enhanced platform durability while reducing weight and manufacturing costs [14].
2. **Dynamic Positioning and Smart Control Systems**
   Dynamic positioning systems use sensors and real-time weather data to optimize turbine orientation and energy capture. Integration with artificial intelligence (AI) further enhances these systems, enabling predictive maintenance and operational efficiency [15].
3. **Hybrid Renewable Energy Solutions**
   Innovative hybrid systems incorporate floating solar panels and battery storage units alongside wind turbines (Fig. 5.11), addressing intermittency issues and maximizing energy output [16]. These hybrid setups promise continuous energy generation and storage solutions that align with grid demands.
4. **Modular and Scalable Designs**
   Modular floating platforms are gaining traction for their cost efficiency and ease of deployment. Prefabricated components can be assembled on-site, reducing transportation costs and installation timelines [13].
5. **Advanced Mooring and Subsea Cabling**
   Mooring systems (Fig. 5.12) are shifting toward synthetic materials like aramid fibers, which offer flexibility and strength in harsh marine conditions. Similarly, high-voltage direct current (HVDC) subsea cables are improving energy trans-

Fig. 5.11 Hybrid wind solar power plant by China's State Power Investment Corporation (SPIC) [17]

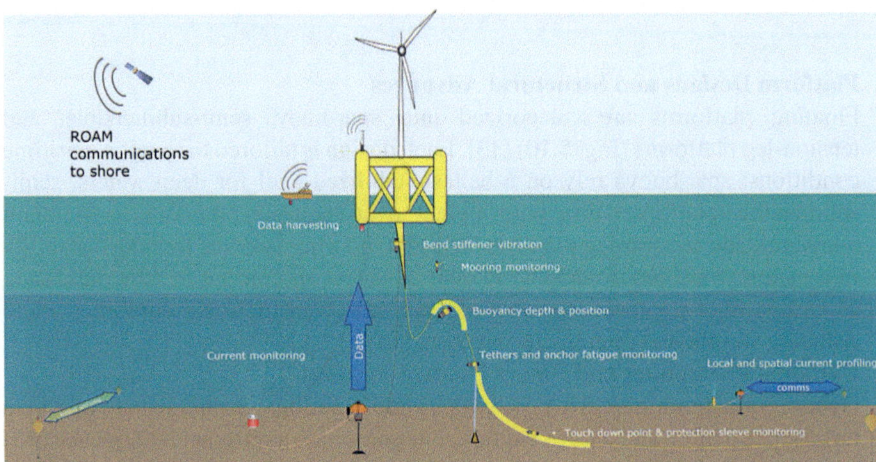

Fig. 5.12 Dynamic cables and mooring systems [18]

mission efficiency, making long-distance offshore wind projects economically viable [14].

5.2.3 The Future of Offshore Wind Energy and Its Economic Implications

Floating offshore wind energy is poised to play a critical role in meeting global renewable energy targets. As nations transition from fossil fuels, this technology offers unparalleled opportunities to harness vast wind resources, particularly in regions with deep coastal waters. However, it is important to recognize that hydrocarbons (oil and gas) will continue to serve as essential transition fuels during this shift. While the depletion of conventional oil and gas resources is inevitable, production declines will likely be gradual, and these fuels will remain integral to energy systems for some time before renewables fully replace them.

1. **Global Market Expansion**
   Countries such as Japan, South Korea, and Norway, with limited shallow-water areas, are investing heavily in floating wind projects. The global market for floating offshore wind is expected to exceed 70 GW by 2040, driven by advancements in technology and government incentives [19].
2. **Economic Growth and Job Creation**
   The floating wind industry is a significant driver of economic growth. It creates jobs across engineering, manufacturing, construction, and maintenance sectors, particularly benefiting coastal communities. For example, projects in Europe alone are expected to generate over 200,000 new jobs by 2030 [20].
3. **Cost Parity with Other Renewables**
   Although floating wind currently involves high capital expenditure, costs are declining due to economies of scale, competitive supply chains, and technological innovation. By 2030, floating wind energy is projected to reach cost parity with fixed-bottom wind and solar power [13].
4. **Environmental and Social Benefits**
   Floating platforms minimize seabed disruption, preserving marine ecosystems and biodiversity. Additionally, their offshore locations reduce visual impact and potential conflicts with fishing and shipping industries [16].
5. **Integration with Green Hydrogen Production**
   Floating wind farms can also support green hydrogen production by powering electrolysis processes directly at sea. This integration addresses energy storage challenges and offers a clean alternative for industrial and transportation sectors [19].

5.2.4 Challenges and Mitigation Strategies

Despite the immense potential of floating offshore wind turbines, their deployment and scalability face several challenges. However, innovative mitigation strategies can address these hurdles effectively.

1. High Capital and Operational Costs
Challenge:
The costs associated with floating wind projects are significantly higher than those of fixed-bottom turbines due to the complexity of platform construction, mooring

systems, and dynamic cabling. Furthermore, operational expenses are amplified by the need for specialized vessels and offshore maintenance teams.
Mitigation Strategies:

- **Economies of Scale:** Increasing project sizes and the adoption of standardized designs can reduce per-unit costs. For instance, collaborative projects across nations can enable shared infrastructure investments [13].
- **Public-Private Partnerships (PPPs):** Governments can incentivize private investment through subsidies, tax benefits, and green bonds, encouraging stakeholders to support floating wind projects.
- **Advances in Materials and Manufacturing:** Utilizing modular components and cost-effective materials like recycled composites can significantly lower production costs.

2. Logistical Complexities in Deployment and Maintenance
Challenge:
Deploying floating turbines involves specialized equipment and port facilities capable of assembling and launching large platforms. Harsh oceanic conditions add to the difficulty of transporting components and conducting regular maintenance.
Mitigation Strategies:

- **Strategic Port Development:** Investments in large-scale offshore wind hubs with assembly, storage, and servicing capabilities can streamline logistics [16].
- **Remote Monitoring and Robotics:** Implementing autonomous drones and underwater robots for inspections and minor repairs minimizes the need for human intervention in dangerous conditions [15].
- **Floating Maintenance Platforms:** The development of mobile, floating maintenance hubs reduces travel time and costs for offshore repairs.

3. Regulatory and Permitting Challenges
Challenge:
Navigating the regulatory frameworks governing marine environments can delay project approvals. Disparate permitting processes and lengthy environmental impact assessments (EIAs)[1] often hinder progress.
Mitigation Strategies:

- **Streamlined Permitting Processes:** Governments can establish unified frameworks and fast-track approvals for renewable energy projects, reducing administrative burdens.
- **Harmonized International Standards:** International collaboration to create standardized guidelines for floating wind projects can simplify cross-border developments [19].

---

[1] Environmental Impact Assessment (EIA) is a formal process that evaluates the potential environmental and social effects of renewable energy projects before implementation, not just a survey but a comprehensive regulatory evaluation.

- **Community Engagement:** Early involvement of local communities in project planning builds trust and mitigates opposition during permitting.

4. Grid Integration and Energy Transmission
Challenge:
Efficiently transmitting electricity from floating turbines to onshore grids remains challenging due to the distance from shore and variable ocean conditions. Losses during energy transfer further reduce economic viability.
Mitigation Strategies:

- **High-Voltage Direct Current (HVDC) Systems:** Advanced subsea cabling with HVDC technology ensures minimal energy loss during transmission [14].
- **Offshore Energy Storage:** Integrating battery systems and hydrogen production facilities with floating wind farms reduces reliance on immediate grid connectivity and stores excess energy.
- **Smart Grid Solutions:** Leveraging AI and machine learning enables dynamic load balancing, ensuring grid stability and reducing downtime [15].

5. Environmental and Ecological Concerns
Challenge:
Floating platforms may disrupt marine habitats, alter local ecosystems, and affect fishing industries. Acoustic disturbances from construction and operation can impact marine life.
Mitigation Strategies:

- **Eco-Friendly Designs:** Developing platforms that provide artificial reefs for marine life can enhance biodiversity rather than harm it.
- **Monitoring and Impact Assessments:** Conducting comprehensive environmental studies before and during operations ensures minimal disruption to ecosystems.
- **Collaborative Solutions with Fisheries:** Engaging with the fishing industry to delineate no-impact zones fosters cooperation and reduces conflicts [16].

6. Technological Maturity and Workforce Expertise
Challenge:
The technology for floating offshore wind is still maturing, with gaps in workforce expertise for installation, operation, and maintenance.
Mitigation Strategies:

- **Research and Development (R&D):** Increased funding for R&D can address technical bottlenecks, including improved mooring systems and more efficient turbine designs.
- **Training Programs:** Establishing specialized training initiatives for engineers, technicians, and operators ensures a skilled workforce [20].
- **Cross-Sector Collaboration:** Lessons from oil and gas offshore operations can be adapted to enhance floating wind technologies.

7. Weather and Climate Risks

Challenge:
Floating wind turbines must withstand extreme weather events, including hurricanes, typhoons, and rough seas, which pose risks to structural integrity and operational efficiency.
Mitigation Strategies:

- **Resilient Design Standards:** Turbine platforms can be engineered to endure extreme conditions through reinforced structures and flexible mooring systems [14].
- **Real-Time Weather Forecasting:** Advanced meteorological tools can predict severe weather, enabling preventive measures such as shutting down turbines or adjusting moorings.Effective deployment of these tools requires consideration of current policies that affect the availability and sharing of meteorological data to ensure forecasting accuracy and operational reliability.
- **Floating Redundancy Systems:** Creating multiple power modules on a single platform ensures that energy production continues even during partial outages.

5.2.4.1 Advanced Aerodynamics and Materials

The evolution of floating offshore wind turbines is intrinsically tied to advancements in aerodynamics and materials science. Innovations in these areas are pivotal for reducing costs, enhancing energy capture efficiency, and ensuring the structural integrity of turbines operating in the challenging offshore environment.

1. Innovations in Aerodynamic Design
Modern turbine blade designs are focused on maximizing energy capture while minimizing drag and turbulence. This is achieved through advanced computational fluid dynamics (CFD) simulations and wind tunnel testing, which enable precise optimization of blade geometries.

- **Longer and Lighter Blades:**
  New aerodynamic profiles allow blades to be longer without becoming excessively heavy, increasing the swept area and improving energy capture. For instance, the development of segmented blades facilitates transportation and installation while maintaining aerodynamic efficiency [21].
- **Smart Blades with Sensors:**
  Embedded sensors monitor wind speed, pressure distribution, and structural strain in real time. This data enables adaptive control systems to adjust the blade pitch dynamically, optimizing performance under varying wind conditions.
- **Tubercle-Inspired Designs:**
  Biomimetic designs inspired by humpback whale fins, featuring tubercles (bumps) along the leading edge, reduce drag and improve lift-to-drag ratios. These designs have shown promise in improving efficiency in turbulent wind environments typical of offshore settings.

2. Advanced Materials for Blades and Structures

The materials used in turbine blades and platforms play a critical role in determining overall performance, durability, and cost-effectiveness. Recent breakthroughs have focused on lightweight, high-strength composites and sustainable alternatives.

- **Carbon Fiber Reinforced Polymers (CFRP):**
  CFRP is increasingly used in blade manufacturing due to its high strength-to-weight ratio and fatigue resistance. These properties are essential for withstanding the high wind loads and saltwater corrosion experienced offshore.
- **Thermoplastic Composites:**
  Unlike traditional thermoset materials, thermoplastics are recyclable, making them an environmentally friendly choice for blade production. Their ease of molding also reduces manufacturing time and costs.
- **Self-Healing Materials:**
  Polymers capable of autonomously repairing small cracks and damage extend the service life of blades and reduce maintenance costs. This technology is particularly valuable in remote offshore locations where repairs can be logistically challenging.
- **Hybrid Platforms:**
  Floating platforms are now constructed using hybrid materials, such as steel-concrete composites, which offer a balance of strength, buoyancy, and cost efficiency. These materials are also designed to resist marine biofouling and corrosion.

3. Integration of Aerodynamics and Materials Science

The integration of advanced materials and aerodynamics has led to synergistic improvements in turbine performance:

- **Lightweight blades with advanced profiles** reduce fatigue on turbine bearings and shafts, enhancing reliability.
- **Aerodynamic coatings** reduce surface roughness and mitigate the impact of environmental degradation, such as erosion from salt spray or sand particles.
- **Smart materials** with shape-memory properties allow components to adapt their form in response to environmental changes, optimizing aerodynamic performance.

4. Cost Reduction Through Innovation

Advanced aerodynamics and materials directly contribute to the economic viability of floating offshore wind projects by addressing key cost drivers:

- **Increased Energy Yield:** Aerodynamically optimized blades improve annual energy production (AEP) by capturing more wind energy, even in low-speed or turbulent conditions.
- **Reduced Maintenance Costs:** Durable materials and smart self-monitoring systems lower the frequency and cost of offshore maintenance.

- **Extended Lifespan:** Enhanced fatigue resistance and environmental durability increase the operational lifespan of turbines, further lowering levelized cost of energy (LCOE).

5. Future Prospects
The future of advanced aerodynamics and materials lies in ongoing research and collaboration between industries and academia:

- **Multi-Material Manufacturing:** The ability to combine materials with distinct properties within a single component will enable further optimization of performance and cost.
- **3D Printing of Components:** Additive manufacturing techniques will allow for rapid prototyping and production of complex aerodynamic designs with minimal waste.
- **AI-Driven Design:** Machine learning algorithms are increasingly being used to model and predict optimal aerodynamic and material properties for new turbine designs.

By combining state-of-the-art aerodynamics with cutting-edge materials science, floating offshore wind turbines are poised to deliver unprecedented levels of efficiency and sustainability. These advancements not only promise lower costs but also position offshore wind as a cornerstone of the global transition to renewable energy.

5.3 Innovations in Battery Storage

As the deployment of renewable energy technologies accelerates, battery storage systems are becoming integral to ensuring energy reliability and grid stability. Innovations in battery storage, particularly in solid-state battery technologies, hold the potential to revolutionize energy storage for both grid and large-scale applications.

5.3.1 Solid-State Battery Technologies

Solid-state batteries (SSB) represent a paradigm shift from conventional lithium-ion batteries (LIB), utilizing solid electrolytes instead of liquid or gel-based ones. This transition offers significant benefits in terms of energy density, safety, and longevity, making them a promising solution for renewable energy storage. A cross sectional view of LIB and SSB is shown in Fig. 5.13.

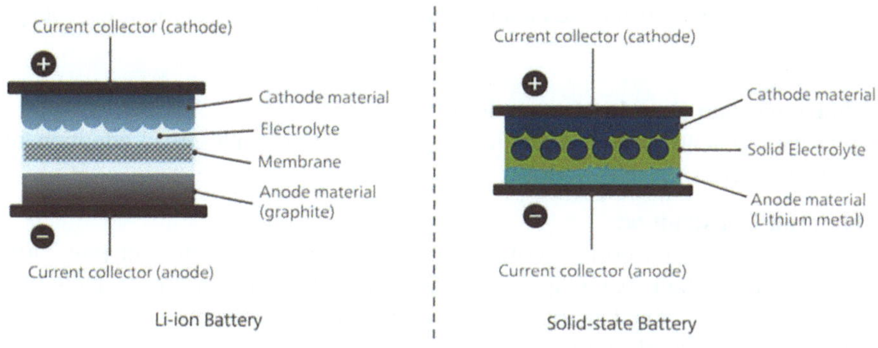

Fig. 5.13 (L) Cross-sectional illustration of a LIB. (R) Cross-sectional illustration of a SSB

1. Role in Improving Energy Density and Safety

- **Higher Energy Density:**
  Solid-state batteries have a greater energy density than traditional lithium-ion batteries, enabling them to store more energy in a smaller space. This is achieved through the use of advanced cathode materials, such as lithium metal anodes, which allow for increased storage capacity. High energy density is particularly critical for grid-scale applications where space constraints can be significant.
- **Enhanced Safety:**
  Unlike liquid electrolytes, solid electrolytes are non-flammable, reducing the risk of thermal runaway and fires. This inherent safety advantage addresses one of the most pressing concerns of large-scale battery installations.
- **Thermal and Chemical Stability:**
  Solid-state batteries exhibit improved thermal stability, allowing them to operate efficiently across a wider temperature range. Additionally, their resistance to chemical degradation extends their operational lifespan, reducing maintenance and replacement costs.

2. Potential for Large-Scale Storage

- **Grid-Level Applications:**
  Solid-state batteries can stabilize intermittent renewable energy sources like offshore wind by storing surplus energy during periods of high generation and discharging it during demand peaks. Their high energy density and extended cycle life make them ideal for such use cases.
- **Scaling Up for Utility Storage:**
  Advances in manufacturing techniques, such as roll-to-roll processing and automated assembly lines, are driving down the cost of solid-state batteries, making them feasible for large-scale deployment. Companies like QuantumScape and Solid Power are pioneering scalable solutions tailored for utility-scale storage.
- **Hybrid Systems Integration:**
  Solid-state batteries can be integrated with other energy storage technologies, such as pumped hydro or flow batteries, to create hybrid systems. These systems

optimize energy delivery and enhance the overall efficiency of renewable energy projects.

3. Challenges and Ongoing Innovations

Despite their potential, several challenges remain in the widespread adoption of solid-state batteries:

- **Material Constraints:**
  The availability and cost of critical materials, such as lithium and rare earth metals, pose a challenge. Ongoing research into alternative materials, including sodium-ion solid-state batteries, aims to address these issues.
- **Manufacturing Scalability:**
  Scaling up production while maintaining quality and performance consistency is a critical hurdle. Collaborative efforts between industry and academia are focusing on streamlining manufacturing processes to meet industrial-scale demand.
- **Durability under Load:**
  Solid-state batteries must withstand high charging and discharging cycles without significant degradation. Innovations in electrode-electrolyte interfaces, such as incorporating flexible materials, are being explored to enhance durability.

4. Future Potential

The integration of solid-state batteries into renewable energy systems is poised to redefine how energy is stored and utilized:

- **Offshore Wind Integration:**
  The compact size and high efficiency of solid-state batteries make them well-suited for floating offshore wind platforms, where space and weight constraints are critical. They could serve as local storage units, stabilizing energy output before transmission to the grid.
- **Decentralized Energy Systems:**
  Solid-state batteries can play a key role in decentralized energy storage solutions, enabling community-level microgrids and off-grid renewable energy projects.
- **Electric Vehicles and Energy Ecosystems:**
  As the automotive industry adopts solid-state technology, these batteries could serve dual purposes: powering vehicles and functioning as part of a larger grid-connected storage network.

In the context of renewable energy, the transformative potential of solid-state batteries cannot be overstated. By addressing the challenges of energy density, safety, and scalability, these innovations are setting the stage for a sustainable and resilient energy future.

5.3.2 Long-Duration Energy Storage

The growing reliance on renewable energy sources, such as solar and wind, has created an increasing need for advanced energy storage solutions that go beyond the traditional 4–6-hour duration. Long-duration energy storage (LDES) technologies are crucial for addressing grid stability challenges and meeting energy demands during peak load periods. By extending storage durations, these technologies enable better integration of intermittent renewable energy, ensuring a reliable and flexible power supply. This section explores the innovations in long-duration energy storage and its significant applications in renewable energy, particularly in grid stability and peak load management.

5.3.2.1 Innovations for Extending Storage Durations

Flow batteries have emerged as one of the most promising solutions for long-duration energy storage. These batteries, such as vanadium redox flow batteries (VRFBs), separate the energy storage and power generation components. This unique design allows for independent scaling of power and storage capacity, making them ideal for applications requiring storage durations of 10 hours or more. One of the key advantages of flow batteries is their durability, with lifespans exceeding 20 years, making them a cost-effective solution for long-term energy storage [22].

Compressed Air Energy Storage (CAES) is another innovative solution capable of providing long-duration storage. CAES systems work by compressing air into underground caverns or containers during periods of low demand and releasing it to drive turbines during peak periods. These systems are capable of storing energy for several days, making them especially useful for balancing seasonal fluctuations in renewable energy generation. Advancements in thermal management have also improved the efficiency of CAES, making it an increasingly viable solution for large-scale energy storage [10].

Hydrogen storage represents a particularly versatile form of long-duration storage. Hydrogen can be produced through electrolysis, using surplus renewable energy, and stored for extended periods, ranging from weeks to months. As an energy carrier, hydrogen can be used for electricity generation, heating, or even industrial processes, offering flexibility in its applications. The integration of hydrogen production with renewable energy systems also enables better management of excess renewable energy, helping to address issues of intermittency and excess supply during off-peak periods [23].

Gravity-based energy storage systems, including pumped hydro and mechanical lifting systems, utilize gravitational potential energy to store and release electricity. Pumped hydro storage has been in use for decades and currently accounts for the majority of long-duration storage capacity worldwide. However, newer systems, like those developed by Energy Vault, use cranes and weights to achieve similar effects in areas where natural water reservoirs are not available. These innovative designs expand the potential for gravity-based storage, particularly in regions where pumped hydro is not feasible [23].

Thermal energy storage is another growing field within long-duration storage, where energy is stored in the form of heat or cold for use over extended periods. Innovations in phase-change materials have increased the efficiency of thermal storage systems, enabling better energy retention and higher storage capacities. Thermal storage is particularly useful for industrial applications where heating or cooling is needed over extended durations, and it integrates seamlessly with existing industrial processes, optimizing energy use [10].

5.3.2.2 Applications of Long Duration Energy in Renewable Energy

Long-duration energy storage plays a critical role in enhancing grid stability, especially as renewable energy sources like solar and wind become more prevalent. By providing energy storage for extended periods, LDES systems smooth out short-term fluctuations in power generation, ensuring that the grid maintains a stable frequency. These systems are particularly important for frequency regulation, where they help to maintain the balance between supply and demand. Additionally, LDES technologies are capable of providing black-start capabilities, allowing the grid to restart after a power outage without relying on fossil fuel-powered plants [22].

One of the most significant benefits of long-duration energy storage is its role in peak load management. LDES systems enable the storage of surplus renewable energy generated during periods of low demand, which can then be released during peak consumption times. This capability helps mitigate the strain on the grid during high-demand periods and reduces the need for peaking power plants, which are often fossil fuel-based. By shifting energy from low-demand to high-demand periods, LDES enhances grid reliability and lowers overall emissions from the power sector [23].

Long-duration energy storage also helps to address the challenges associated with renewable energy intermittency. By pairing LDES systems with renewable energy projects, such as offshore wind farms or large-scale solar installations, the dispatchability of renewable power is significantly increased. This integration ensures that renewable energy can be reliably delivered even when the wind is not blowing or the sun is not shining. The ability to store energy over long periods reduces the need for curtailment, which occurs when renewable energy generation exceeds grid demand and results in wasted energy. By minimizing curtailment, LDES systems improve the overall economics of renewable energy projects and ensure a more stable energy supply [10].

In addition to improving grid stability and peak load management, long-duration energy storage technologies enable greater integration of renewable energy into the broader energy mix. Hybrid systems that combine renewable generation with long-duration storage are becoming increasingly popular. For example, pairing solar or wind farms with hydrogen storage systems can help address seasonal fluctuations in renewable generation, providing a continuous supply of energy regardless of seasonal variations. This synergy between renewable energy production and long-duration storage technologies ensures a more resilient and sustainable energy grid [22].

5.3.2.3 Challenges and Future Directions

While the potential of long-duration energy storage is vast, several challenges remain to be addressed. A key barrier to widespread adoption is the high initial cost of technologies such as hydrogen storage, compressed air energy storage (CAES), and advanced flow batteries. These systems often require substantial capital investment, which can deter energy providers. However, not all long-duration storage technologies are equally speculative. For instance, flow batteries and hydrogen storage have moved beyond early-stage research, with pilot and demonstration projects indicating growing technological maturity and cost-reduction potential. In contrast, other concepts like certain novel CAES configurations remain largely theoretical or in the experimental phase. As research and development efforts continue, particularly for technologies with a higher technology readiness level, costs are expected to decline through economies of scale, improved system designs, and advances in manufacturing processes [23].

Infrastructure requirements also pose a challenge, particularly for technologies like CAES and gravity-based storage, which require specific geographic features or large amounts of land. While pumped hydro storage has been in use for many years, its deployment is limited to areas with suitable water reservoirs, and the same is true for some gravity-based systems. As a result, the availability of suitable locations for these storage systems may be a limiting factor in their widespread deployment [10].

Policy support and market incentives will play a crucial role in the adoption of long-duration energy storage technologies. However, one challenge lies in the current policy environment, where shifting priorities and administrative changes can create uncertainty for investors and developers. While some federal and state programs promote energy storage, others lack the clarity or consistency needed to fully support long-duration solutions. To ensure meaningful progress, governments and regulatory bodies must commit to stable, long-term frameworks that reduce barriers, incentivize innovation, and integrate storage technologies into broader decarbonization strategies [22]. Clear, durable policy signals will be vital for building confidence in this emerging sector and achieving a low-carbon, resilient energy system.

Despite these challenges, ongoing research and development are rapidly addressing the barriers to long-duration energy storage. Advancements in materials science, manufacturing techniques, and system design are driving down costs by up to 25% and improving efficiency by approximately 15% across key long-duration storage technologies.Moreover, as the demand for renewable energy grows and the need for energy storage becomes more critical, the adoption of LDES solutions will become increasingly viable and essential for ensuring a sustainable and reliable energy future [23].

In conclusion, Long-duration energy storage is poised to play a transformative role in the energy transition, providing a critical solution for grid stability, peak load management, and the integration of renewable energy. Through innovations such as flow batteries, CAES, hydrogen storage, and gravity-based systems, LDES technologies offer a path to a more reliable, resilient, and sustainable energy grid. Although challenges remain in terms of cost, infrastructure, and policy support, the continued advancement of LDES technologies promises to enable the widespread deployment of renewable energy and drive the transition to a low-carbon future.

5.3.3 Flow Batteries and Alternative Storage Solutions

As the global energy transition accelerates, the demand for large-scale, long-duration energy storage solutions has become more pronounced. While conventional battery technologies such as lithium-ion are prevalent in applications requiring short-term storage, they face limitations when it comes to handling the prolonged storage needs of renewable energy systems. Flow batteries have emerged as a promising solution to address these challenges, offering substantial advantages in terms of scalability, energy duration, and cost-effectiveness. This section delves into the potential of flow batteries for large-scale, long-term energy storage, alongside an exploration of alternative storage solutions such as compressed air and thermal storage, which are gaining traction for their ability to address the evolving needs of modern energy grids.

5.3.3.1 Exploration of Flow Battery Technologies for Large-Scale, Long-Term Energy Storage

Flow batteries are an innovative class of electrochemical storage technologies that offer distinct advantages over traditional batteries. Unlike conventional batteries, where energy is stored within the electrode material, flow batteries store energy in two liquid electrolyte solutions that are kept in separate tanks. These electrolytes flow through a reactor cell, where electrochemical reactions take place to charge or discharge the battery. This architecture allows for energy and power to be scaled independently, which is particularly valuable for applications requiring long-duration energy storage [24].

Among the various types of flow batteries, the **vanadium redox flow battery (VRFB)** is one of the most well-known and widely researched technologies. VRFBs use vanadium ions in both the positive and negative electrolytes, which prevents the issues of cross-contamination that typically occur in other types of flow batteries. This unique design not only enhances the cycle life of the battery but also allows it to maintain high efficiency over extended periods. The modular nature of VRFBs makes them particularly suitable for large-scale, long-term storage applications, such as grid stabilization, peak shaving, and renewable energy integration. VRFBs are capable of providing storage durations of 10 hours or more, making them a viable option for balancing the intermittent nature of renewable energy sources like wind and solar [25].

One of the primary benefits of flow batteries is their long cycle life. Flow batteries can be cycled tens of thousands of times without significant degradation, making them ideal for applications that demand long-term energy storage. Additionally, flow batteries are capable of storing energy for extended periods without suffering from the self-discharge issues that limit traditional battery technologies. This makes them particularly well-suited for seasonal energy storage, where energy generated in one season can be stored and used in another, providing a solution to the intermittency challenges of renewable energy [24].

The scalability and versatility of flow batteries offer a promising pathway to reducing the costs of energy storage. As the technology matures, economies of scale and advances in materials and manufacturing are projected to reduce capital costs by up to 30% over the next decade. Unlike lithium-ion batteries, flow batteries rely less on rare and costly materials such as lithium and cobalt, which helps lower raw material expenses and improve sustainability. For context, in markets like California, utilities have reported that the levelized cost of electricity (LCOE) from renewables plus storage systems is now competitive with or lower than that from traditional fossil fuels, often cited around $20–40 per megawatt-hour (MWh) for solar and wind, compared to $40–70 per MWh for natural gas plants [26, 27]. These cost dynamics reinforce the growing economic viability of flow batteries as a large-scale storage solution supporting aggressive renewable energy targets [25].

However, flow batteries still face challenges related to efficiency and energy density. Compared to lithium-ion batteries, flow batteries have lower energy densities, requiring larger physical volumes to store the same amount of energy (see Table 5.1). This can pose logistical challenges, especially in urban areas. Table 5.1 summarizes the key characteristics of several long-duration energy storage technologies, including deployment status, costs, energy density, and efficiency.

As shown in Table 5.1, flow batteries have lower energy densities compared to lithium-ion batteries, requiring larger physical volumes to store the same amount of energy. This can pose logistical challenges, especially in densely populated areas.

Table 5.1 Comparison of long-duration energy storage technologies

Technology	Deployment status	Levelized cost ($/kWh)	Energy density (Wh/L)	Efficiency (%)
Flow batteries	Early commercial	150–250	20–50	65–75
Lithium-ion	Mature	100–150	250–700	85–95
Hydrogen storage	Pilot/Demo	200–300	~1000	30–40
CAES	Limited deployment	50–150	~1000	40–50

5.3.3.2 The Potential of Other Emerging Storage Solutions

While flow batteries are a promising option for long-duration energy storage, other emerging technologies also offer potential for large-scale energy storage. Among these, compressed air energy storage (CAES) and thermal energy storage are gaining significant attention due to their ability to provide energy storage for extended durations, often ranging from several hours to days.

Compressed Air Energy Storage (CAES) involves the compression of air into underground caverns or storage vessels during periods of low electricity demand. When energy is needed, the compressed air is released, heated, and expanded through turbines to generate electricity. This technology has been in use for decades and continues to evolve with advancements in efficiency and storage duration. The primary advantage of CAES is its ability to store large amounts of energy, making it ideal for grid-scale applications. Furthermore, CAES can provide long-duration storage for up to several days, which is crucial for balancing the intermittency of renewable energy sources [23].

The potential of CAES has been significantly enhanced by recent advancements in adiabatic CAES (A-CAES), where heat generated during the compression process is stored and reused during the expansion phase. This increases the efficiency of the system and reduces the need for external heating sources. A-CAES systems have the ability to store energy for longer periods while maintaining higher efficiency, making them a more attractive solution for long-duration storage applications (IEA, 2021). One of the key challenges for CAES systems, however, is the availability of suitable geological formations, such as salt caverns, which are needed for energy storage. Despite this limitation, CAES remains a promising technology, particularly in regions with abundant underground storage potential.

Thermal Energy Storage (TES) involves storing heat in a medium, such as molten salt, for later use in power generation or industrial applications. One of the most widely used forms of TES is in concentrated solar power (CSP) plants, where excess solar energy is stored as heat in molten salt tanks during the day and then used to generate electricity during the night or on cloudy days. The advantage of TES systems is their high efficiency in energy retention and the ability to store energy over long periods without significant losses. Recent advancements in phase-change materials (PCMs) have improved the energy storage capabilities of thermal systems, allowing for better energy density and greater storage capacity. TES is particularly useful in industrial applications that require heat or cooling, such as in the manufacturing or chemical industries, where it can serve as a reliable backup energy source [10].

Additionally, **cryogenic energy storage**, which involves storing energy in the form of supercooled liquid air, is an emerging alternative to traditional energy storage solutions. In this process, excess energy is used to liquefy air, which is then stored in insulated containers. When energy is needed, the liquid air is evaporated and expanded to drive turbines, generating electricity. This technology has the potential for long-duration storage, with applications in both grid-scale and remote energy storage. One of the main advantages of cryogenic energy storage is its ability to be deployed in areas with limited access to other forms of energy storage infrastructure [23].

In conclusion, Flow batteries and alternative energy storage solutions, such as compressed air energy storage and thermal energy storage, are emerging as critical technologies for large-scale, long-duration energy storage. Flow batteries, with their modular design and long cycle life, offer a scalable and cost-effective solution for integrating renewable energy sources into the grid and addressing the intermittency challenges. Meanwhile, technologies like CAES, thermal storage, and cryogenic storage provide complementary solutions that further enhance grid stability and energy reliability. As these technologies continue to evolve and improve, they will play an essential role in enabling the transition to a more sustainable and resilient energy future.

5.4 Geothermal Energy

Geothermal energy refers to the immense heat stored within the Earth's interior, which has three primary sources. The first is primordial heat, a remnant from the planet's initial formation and differentiation, which has remained trapped beneath the surface for billions of years. The second source is the radioactive decay of isotopes such as uranium, thorium, and potassium, which continually generate heat within the Earth's crust and mantle. The third is frictional heating, which results from the movement of molten material in the mantle and outer core, often influenced by the gravitational pull of the Moon. Collectively, these processes create internal temperatures that exceed 5200 °C, making Earth's core even hotter than the surface of the sun. As a result, the Earth emits approximately 47 TW of heat into space, representing a vast and underutilized energy resource.

Depending on the temperature and depth of this heat, geothermal energy can serve several applications. It can generate electricity through steam turbines, directly heat buildings through radiators or district heating systems, or provide steady ground temperature for geothermal heat pumps used in residential heating and cooling. Each use case requires specific subsurface conditions to be efficient. In particular, regions with a high geothermal gradient and accessible hot rock formations are ideal for geothermal exploitation. However, these regions are often tectonically active and require careful assessment of seismic risks and drilling feasibility.

To effectively extract geothermal energy for power generation, three main conditions must be met: there must be sufficiently hot rocks relatively close to the surface, fluids such as water or brine must be available to absorb and transport the heat upward, and there must be permeability within the rock to allow the heated fluid to circulate and rise (Fig. 5.14). These conditions are most commonly found in volcanically and tectonically active regions. Examples include Kilauea in Hawai'i, which is heated by upwelling mantle plumes, and Yellowstone National Park, which lies atop a shallow magma intrusion. Other high-potential regions include Iceland,

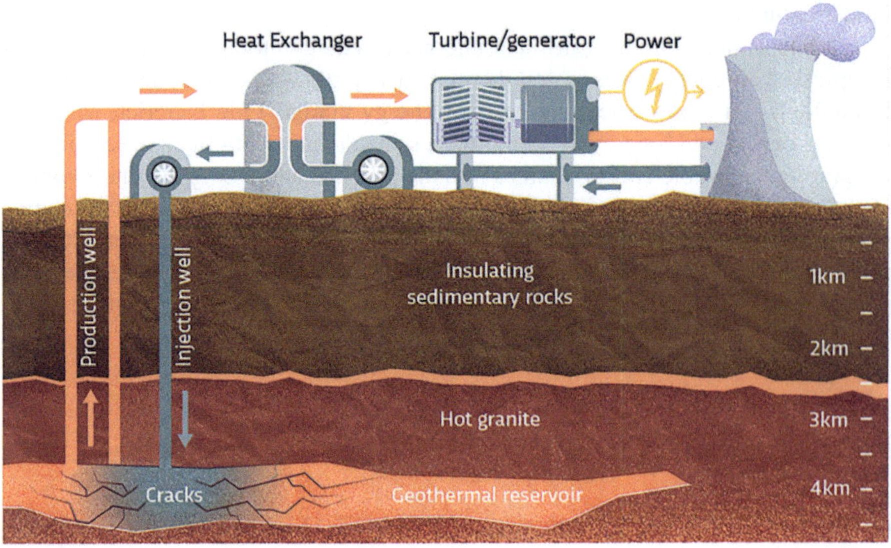

Fig. 5.14 Conventional Geothermal power plant [28]

New Zealand, and parts of California such as the Geysers geothermal field. These areas provide hydrothermal systems where naturally circulating groundwater is heated by nearby magma bodies, resulting in geysers and hot springs. However, these ideal geological conditions are relatively rare and not always located near centers of energy demand.

Geothermal resources are generally classified by temperature. High-temperature resources, exceeding 180 °C, are suitable for generating live steam to drive turbines directly. Medium-temperature resources between 70 and 180 °C are often used for deep direct use (DDU), such as heating large buildings or greenhouse agriculture. Low-temperature systems under 70 °C are commonly used in residential ground-source heat pumps, which exploit the Earth's stable shallow temperature (typically 10–20 °C) to reduce the energy required for heating and cooling buildings. In some cases, depleted oilfields can be repurposed for geothermal use, particularly when existing wellbores can be used to avoid the costs of new drilling.

A significant operational challenge in geothermal systems is mineral scaling. As geothermal fluids cool during extraction, dissolved minerals such as calcite and quartz can precipitate and form solid deposits in pipes and heat exchangers. These blockages reduce system efficiency and require regular maintenance. The Rotokawa Power Plant in New Zealand, for example, is well known for the rapid accumulation of scale in its surface equipment. In that case, however, the mineral deposits contain valuable concentrations of gold, which help offset the cost of regular pipe cleaning and component replacement.

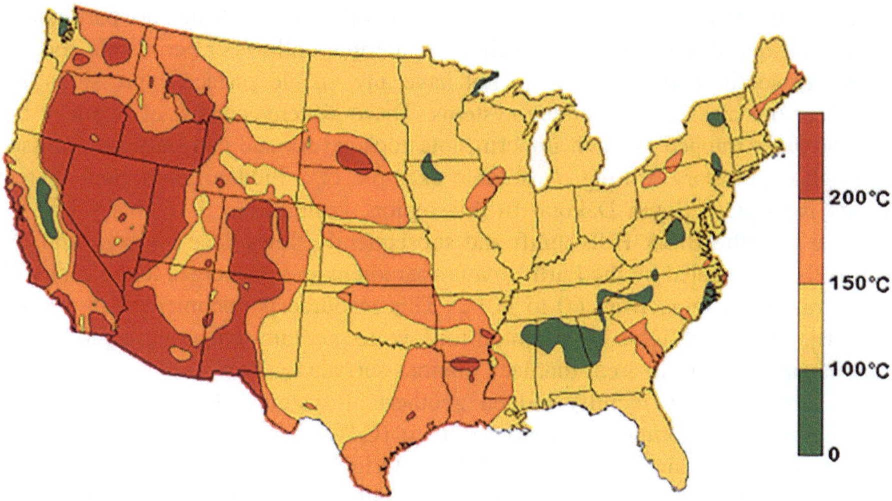

Fig. 5.15 Geothermal gradient map of the U.S. [29]

5.4.1 Geothermal Resource Evaluation

Geothermal development has historically focused on regions where surface expressions of heat are evident, such as hot springs, geysers, fumaroles, and steam vents. These features are typically located along tectonic boundaries, especially within the Pacific Ring of Fire. In the United States, geothermal electricity generation accounts for less than 0.5% of the national energy supply and is largely confined to the West Coast, particularly in California near the San Andreas Fault and the Cascadia Subduction Zone.Below is a gradient map of the U.S. (Fig. 5.15)

In contrast, countries with more active volcanic geology have developed geothermal energy on a much broader scale. For instance, Iceland sources approximately 25% of its electricity from geothermal energy and uses it to heat over 85% of its homes and buildings. Kenya, situated along the East African Rift, generates nearly half of its electricity from geothermal resources. Indonesia and New Zealand also utilize geothermal systems extensively.

A key metric in resource evaluation is the geothermal gradient, the rate at which temperature increases with depth. On average, this gradient ranges from 25 to 30 °C per kilometer of depth. In regions with steep gradients, sufficient temperatures for electricity generation (typically above 200 °C) can be reached at depths of 3–4 km. In areas with shallower gradients, drilling must go deeper, which can increase project costs and complexity. To estimate subsurface temperatures, geoscientists often use bottom hole temperature (BHT) data collected from oil and gas wells. However, BHT readings can be distorted by fluid movements within the wellbore, such as cool groundwater infiltrating from above or hot gases rising from below. For more accurate evaluations, continuous temperature logs are preferred. These logs measure temperature along the entire depth of the borehole and are particularly useful for identifying anomalies, such as geothermal upflows or permeable zones.

Many geothermal resources do not have any visible surface expressions and are therefore classified as hidden systems. These systems consist of warm or hot groundwater trapped beneath impermeable rock layers and can only be accessed through exploratory drilling. A notable example of a hidden geothermal system exists in western South Dakota. In this region, groundwater is recharged at the surface near the Black Hills uplift and subsequently flows deep underground. As it travels, it is heated by the Earth's natural gradient and reemerges to the east with temperatures ranging from 60 to 80 °C. This resource is documented in a study by [30], which highlights the potential for low- to medium-temperature geothermal heating applications in areas that lack obvious surface features.

To systematically identify hidden geothermal systems, the U.S. Department of Energy's Geothermal Technology Office (GTO) has adapted a methodology known as Play Fairway Analysis (PFA). Originally developed for petroleum exploration, PFA involves analyzing geological, hydrological, and thermal data to delineate regions with the highest likelihood of viable geothermal resources. This approach has successfully been applied in states such as Nevada, Idaho, Washington, and Hawai'i. In addition to PFA, the GTO supports a variety of geophysical imaging technologies, including gravity surveys, seismic reflection profiling, and magnetotelluric (MT) surveys, which help visualize the subsurface structure and identify promising reservoirs. These tools are typically combined with surface geological mapping and geochemical sampling to refine exploration targets and reduce drilling risks.

5.4.2 *Types of Geothermal Power Plants*

There are three primary types of geothermal power plants used for electricity generation, each designed to accommodate specific temperature ranges and fluid characteristics.

The first type is the direct dry steam plant (Fig. 5.16), which utilizes steam that emerges from geothermal wells at very high temperatures, typically between 200 and 400 °C. This live steam is routed directly to a turbine that drives an electric generator. The steam must be dry, meaning it contains no liquid water droplets, because even small amounts of moisture can cause erosion and damage to the turbine blades due to their high rotational speed. Dry steam systems are rare and are only viable in regions with exceptionally hot and permeable reservoirs, such as the Geysers field in California or the Larderello field in Italy.

The second type is the flash steam plant (Fig. 5.17), which uses high-pressure hot water from underground reservoirs that is brought to the surface in a liquid state. When this pressurized fluid is released into a lower-pressure environment, such as a separator tank, it rapidly vaporizes or "flashes" into steam. This steam is then used to power a turbine. In a double-flash system, the remaining liquid after

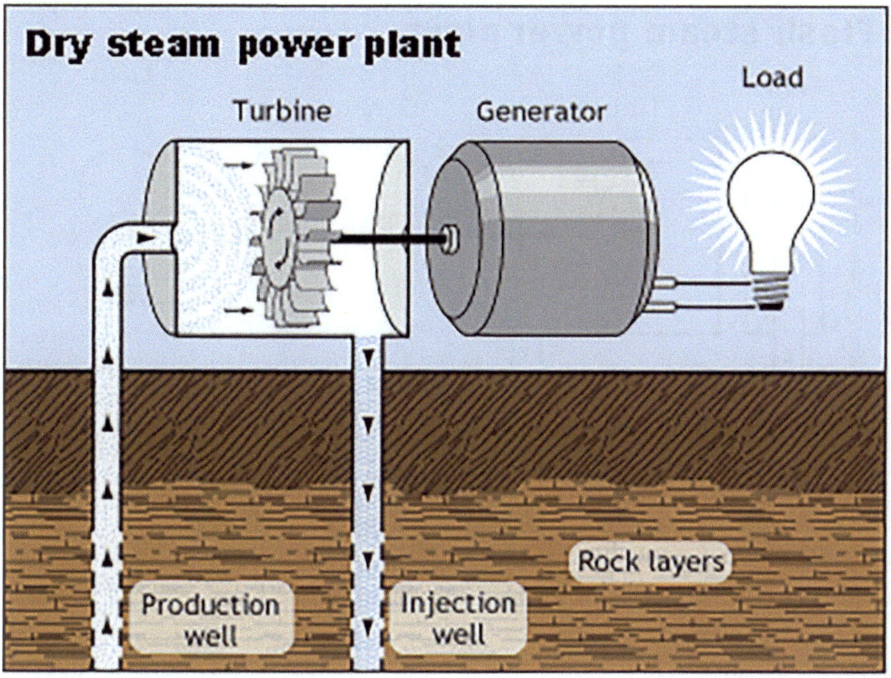

Fig. 5.16 Direct Steam Plant [31]

the first flash is subjected to another pressure drop, generating additional steam and improving efficiency. Flash steam plants are more common than dry steam plants and can operate with slightly lower resource temperatures.

The third and most versatile type of geothermal plant is the binary cycle plant (Fig. 5.18). This system is specifically designed for moderate-temperature resources, typically below 200 °C. In a binary system, the geothermal fluid never directly enters the turbine. Instead, it passes through a heat exchanger where it transfers its heat to a secondary working fluid, usually a hydrocarbon such as isobutane or pentane, that has a lower boiling point than water. The secondary fluid vaporizes, spins the turbine, and is then condensed and cycled back through the system. Binary cycle plants are closed-loop systems, meaning there is no atmospheric release of geothermal fluid or working fluid, making them environmentally clean. Because they can operate at lower temperatures and are scalable to a wide range of geological settings, binary systems are expected to become the dominant form of geothermal power generation in the future.

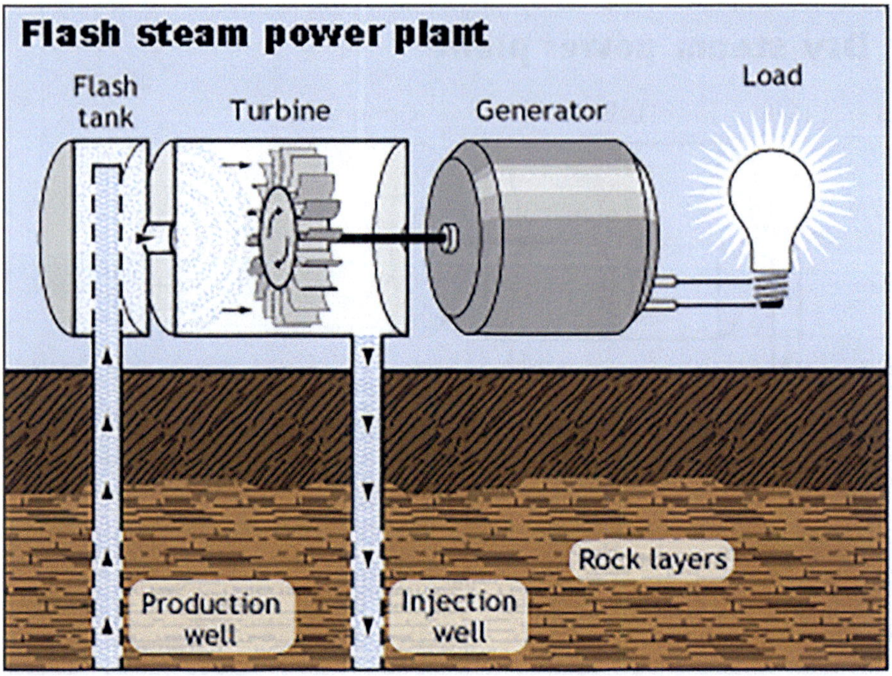

Fig. 5.17 Flash Stream Plant [31]

5.5 Integration with Smart Grids and IoT

The shift towards renewable energy requires robust systems capable of handling the variability of renewable resources while ensuring grid stability, efficiency, and reliability. Smart grids, enhanced by the Internet of Things (IoT), have emerged as transformative technologies, reshaping how energy is generated, distributed, and consumed. This section explores how these advancements optimize energy systems and support a sustainable energy future.

Smart Grids for Optimized Energy Distribution
Smart grids modernize traditional energy systems by integrating digital communication technologies, automation, and advanced analytics. Unlike conventional grids, which rely on rigid, one-directional power flows, smart grids enable two-way energy and information exchange between utilities and consumers. This capability is crucial for renewable energy integration, as it accommodates fluctuations in energy production caused by variable resources like solar and wind.

One of the core advantages of smart grids is their capacity to balance supply and demand dynamically. Through predictive analytics and real-time monitoring, smart grids can anticipate shifts in energy needs and adjust distribution accordingly.

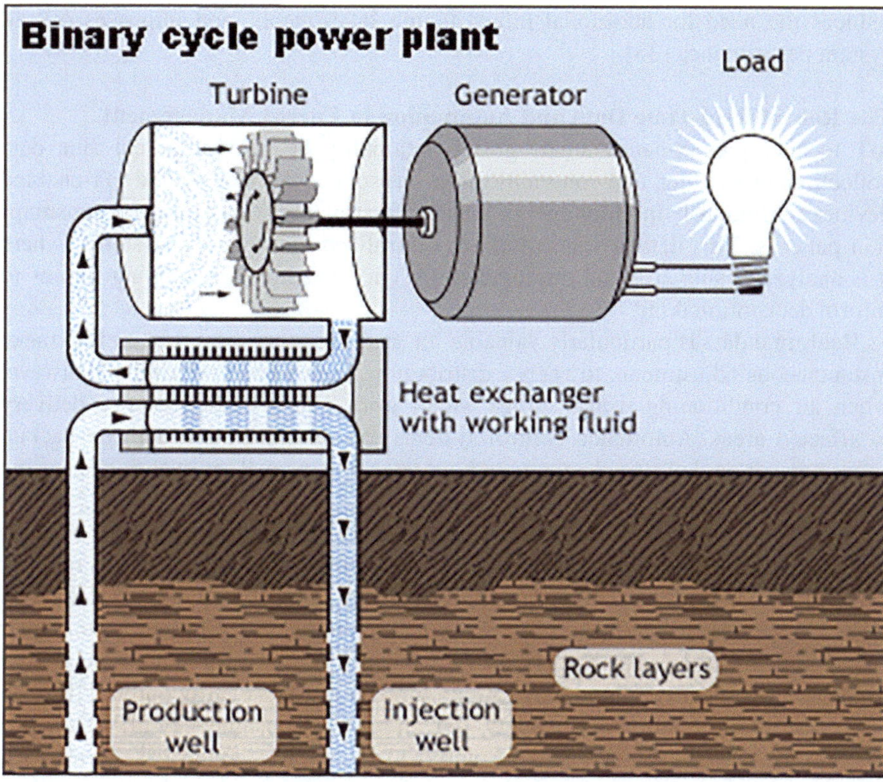

Fig. 5.18 Flash steam power plant [31]

For example, when solar energy production peaks during midday, smart grids can redirect excess power to battery storage systems or export it to other grid sections facing high demand. During periods of low production, such as at night, the stored energy can be utilized to maintain equilibrium.

Incorporating distributed energy resources (DERs), such as rooftop solar panels, wind turbines, and electric vehicles, into the grid is another key feature of smart grids. This decentralized approach reduces dependency on large power plants, minimizes transmission losses, and enhances the grid's overall resilience. By monitoring and managing DERs, smart grids support a cleaner, more diversified energy mix and create opportunities for communities to become energy-independent [32].

Demand response programs, enabled by smart grids, further enhance energy distribution efficiency. These programs allow utilities to adjust consumer energy consumption during peak periods, reducing strain on the grid. For instance, utilities can offer incentives to consumers who delay energy-intensive activities, like charging electric vehicles, to off-peak hours. This practice smooths demand curves, reduces the need for additional infrastructure investments, and improves overall system performance [33].

The Role of Real-Time Data and Automation in Energy Management

IoT technologies enhance smart grid functionality by providing real-time data collection, processing, and communication. Sensors, smart meters, and IoT-enabled devices continuously monitor energy flows, infrastructure conditions, and consumption patterns. This data is transmitted to centralized management systems, where it is analyzed using artificial intelligence (AI) and machine learning algorithms to inform decision-making.

Real-time data is particularly valuable for grid operators, who rely on it to make instantaneous adjustments to energy distribution. For example, during a heatwave, when air conditioning usage surges, smart grids can prioritize energy delivery to affected areas. Automated control systems can reroute power, activate backup generators, or deploy stored energy without requiring manual intervention, ensuring a stable supply during critical periods.

Proactive maintenance is another significant benefit of IoT integration. Sensors installed on critical grid components, such as transformers and substations, monitor performance metrics and detect anomalies. Early detection of potential failures allows operators to perform preventive maintenance, reducing outages, repair costs, and downtime. This capability is particularly crucial for renewable energy infrastructure, where equipment reliability directly impacts energy output [10].

In addition to operational benefits, IoT empowers consumers to play an active role in energy management. Smart home devices, such as connected thermostats, lighting systems, and energy monitors, enable users to track and optimize their energy usage. Consumers can program appliances to operate during off-peak hours, reducing their electricity bills and contributing to grid stability. Moreover, households with renewable installations, such as rooftop solar panels, can sell surplus energy back to the grid, creating a decentralized energy ecosystem that benefits both consumers and utilities.

Towards a Sustainable and Intelligent Energy Ecosystem

The synergy between smart grids and IoT is revolutionizing the energy sector, enabling the seamless integration of renewable energy and creating a more resilient, efficient, and sustainable energy system. These technologies reduce energy waste, enhance grid flexibility, and empower both utilities and consumers with unprecedented control and insights.

As the adoption of renewable energy accelerates, the demand for smart grid infrastructure is expected to grow. Governments and private entities worldwide are investing in smart grid projects to modernize aging infrastructure, accommodate new energy sources, and address climate change. For example, the European Union's Green Deal includes significant funding for smart grid deployment to support its renewable energy goals. Similarly, countries like China and India are incorporating smart grids into their energy strategies to meet the dual objectives of energy access and sustainability.

The future of smart grids lies in their ability to integrate emerging technologies, such as blockchain for secure energy transactions and quantum computing for advanced analytics. These innovations will further enhance grid efficiency, resilience, and scalability, ensuring that energy systems remain robust in the face of growing complexity and demand.

Challenges and Opportunities

Despite their potential, implementing smart grids and IoT-based systems presents challenges. High initial costs, cybersecurity concerns, and the need for regulatory support can hinder widespread adoption. Additionally, integrating diverse energy resources and maintaining grid interoperability requires substantial planning and coordination.

However, the benefits of smart grids far outweigh these challenges. By enabling more efficient energy distribution, enhancing renewable energy integration, and fostering consumer engagement, smart grids represent a critical component of the global transition to a clean energy future. Their integration with IoT amplifies their impact, paving the way for an intelligent, interconnected, and sustainable energy ecosystem.

5.5.1 IoT in Renewable Energy Systems

The Internet of Things (IoT) is revolutionizing renewable energy systems by enabling seamless communication between devices, sensors, and control systems. Through real-time monitoring, predictive analytics, and automated responses, IoT empowers stakeholders to enhance energy generation, optimize storage solutions, and streamline distribution networks. This integration is not only boosting the efficiency of renewable energy systems but also driving their scalability and economic feasibility.

5.5.1.1 IoT for Monitoring and Optimizing Energy Generation, Storage, and Distribution

IoT plays a pivotal role in monitoring renewable energy generation systems, such as solar farms, wind turbines, and hydroelectric plants. Sensors embedded in renewable energy infrastructure continuously gather data on parameters like solar irradiance, wind speed, and turbine performance. This data is transmitted to centralized platforms, where advanced analytics assess system efficiency and detect anomalies. By enabling predictive maintenance, IoT reduces downtime and extends the operational lifespan of critical components [34].

For energy storage systems, IoT offers unparalleled control and visibility. Battery management systems equipped with IoT sensors monitor charge levels, temperature, and overall health, ensuring optimal operation. This capability is especially crucial for large-scale storage solutions like lithium-ion batteries or emerging technologies like solid-state batteries. IoT-enabled systems can dynamically adjust charging and discharging cycles to align with grid demands, thus improving energy utilization and minimizing waste.

IoT also optimizes energy distribution by facilitating demand-side management and load balancing. Smart meters, grid-connected devices, and IoT-enabled substations provide real-time data on electricity consumption patterns. Utilities can analyze this data to predict demand fluctuations and adjust energy distribution accordingly. Moreover, IoT enhances the integration of distributed energy resources (DERs), such as rooftop solar panels and community wind projects, by coordinating their contributions to the grid. This approach reduces transmission losses and enhances grid stability, creating a more resilient energy ecosystem [19].

5.5.2 Expanded Case Studies on IoT-Enabled Renewable Energy Systems

1. Enel Green Power's Smart Wind Farms

Enel Green Power has been at the forefront of integrating IoT technologies into wind energy operations. The company employs IoT-enabled sensors on wind turbines to monitor critical metrics such as rotor speed, wind direction, vibration levels, and temperature. These sensors collect real-time data, which is transmitted to a central monitoring hub powered by advanced analytics and machine learning algorithms.

One notable example is Enel's operations in Spain, where the IoT system allowed the company to implement predictive maintenance strategies. By identifying potential issues such as blade fatigue or gearbox wear early, technicians could address problems before they caused significant downtime. This proactive approach led to a 20% reduction in maintenance costs and a 15% increase in turbine efficiency over a two-year period. Moreover, the integration of IoT with weather forecasting tools allowed Enel to optimize turbine operations during periods of high wind speeds, further increasing energy yield.

2. Tesla Powerwall and Virtual Power Plants (VPPs)

Tesla's Virtual Power Plant (VPP) project in South Australia is a landmark initiative demonstrating the potential of IoT in decentralized energy systems. The VPP aggregates energy stored in thousands of Tesla Powerwall batteries installed in residential homes and small businesses. Each Powerwall unit is equipped with IoT capabilities, enabling real-time communication with the central control system.

The IoT network allows Tesla to monitor battery charge levels, predict energy demand, and coordinate energy dispatch across the VPP. For instance, during a heatwave in January 2023, the VPP successfully supplied 20 MW of power to the grid, preventing blackouts and stabilizing grid operations. The project also reduced electricity bills for participants by up to 30%, showcasing the dual benefits of IoT-enabled systems for individual users and the broader grid.

Additionally, Tesla uses machine learning algorithms to predict energy consumption patterns and dynamically adjust charging and discharging cycles. This ensures that Powerwall batteries are always ready to meet both household needs and grid demand. The project serves as a blueprint for how IoT can support the transition to distributed, renewable energy systems on a global scale.

3. IoT in India's Solar Farms

India has rapidly expanded its solar energy capacity, with IoT playing a crucial role in ensuring the efficiency and reliability of large-scale solar farms. Tata Power Solar, one of the country's leading renewable energy companies, has implemented IoT solutions in its solar farms across Rajasthan and Gujarat.

IoT-enabled sensors monitor parameters such as solar panel temperature, dust accumulation, and inverter performance. In Rajasthan, where dust storms are frequent, IoT sensors provide real-time alerts when panels need cleaning. Automated cleaning systems, triggered by IoT data, have reduced manual labor and improved panel efficiency by 12%.

In addition, IoT platforms integrate with weather forecasting models to predict solar irradiance and optimize energy generation schedules. By combining these forecasts with real-time data from the grid, Tata Power Solar can adjust the farm's output to match grid demand, minimizing curtailment and maximizing revenue.

The success of IoT-enabled solar farms in India has also encouraged the adoption of similar technologies in smaller, decentralized projects, such as microgrids in rural areas. These microgrids use IoT to monitor energy consumption and ensure efficient distribution, bringing reliable electricity to remote communities.

4. Smart Microgrids in Africa

While not tied to a specific corporation, the development of IoT-enabled smart microgrids in Africa highlights the transformative potential of technology in underserved regions. In Kenya, a project led by the company PowerGen Renewable Energy uses IoT to manage microgrids powered by solar and wind energy.

IoT sensors installed in these microgrids monitor energy generation, battery storage levels, and local consumption patterns. This data is transmitted to cloud-based platforms, where algorithms predict energy demand and optimize resource allocation. In one village, the IoT system identified patterns of peak energy usage in the evenings, prompting the microgrid to prioritize battery storage during the day.

This approach reduced outages and increased the reliability of electricity supply for hundreds of households. Additionally, the IoT system allowed users to track their energy usage via mobile apps, encouraging efficient consumption. The project has become a model for how IoT can enable sustainable development and energy access in low-income regions.

5. Siemens Gamesa's Wind Power IoT Solutions

Siemens Gamesa Renewable Energy has leveraged IoT in its offshore and onshore wind farms to optimize performance and reduce operational costs. The company's

IoT platform, powered by artificial intelligence, continuously monitors turbine health and environmental conditions.

For example, at a wind farm in Denmark, Siemens Gamesa used IoT to collect real-time data on wind patterns, turbine vibrations, and component wear. The system's predictive analytics identified potential failures in the gearbox and recommended maintenance before any operational disruptions occurred. This proactive maintenance strategy extended the lifespan of critical components and reduced overall maintenance expenses by 25%.

Moreover, Siemens Gamesa's IoT solutions integrate with blockchain technology to enable secure, transparent energy trading. By tracking energy generation and usage data, the system allows wind farm operators to sell excess power directly to consumers or other businesses, streamlining energy markets and reducing transaction costs.

5.6 Resource Assessment and Optimization

Renewable energy project development relies on **accurate resource assessment** to determine feasibility, optimize siting, and enhance energy output. Advances in the analytical techniques, **geographic information systems (GIS)**, and **machine learning** (ML) have significantly improved **solar and wind resource analysis**, leading to **better site selection and forecasting accuracy**.

5.6.1 Solar Irradiance and Wind Speed Analysis

Accurate analysis of solar irradiance and wind speed is fundamental for assessing the feasibility and optimizing the performance of renewable energy projects. These environmental parameters directly impact the potential energy yield from solar and wind resources. Advances in reliable forecasting and spatial assessment has helped developers identify optimal project sites, reduce uncertainty, and maximize return on investment. The following sections explore the analytical methods used to evaluate these resources, beginning with solar irradiance.

5.6.1.1 Assessing Solar Irradiance

Solar energy generation depends on **solar irradiance**, which varies due to geographical, atmospheric, and seasonal conditions. Analytical methods used in resource assessment include:

1. Satellite and Ground-Based Measurements

- **Satellite Data**: NASA's Surface Meteorology, NREL's National Solar Radiation Database (NSRDB), and Copernicus datasets provide high-resolution irradiance data.
- **Ground-Based Measurements**: Pyranometers and pyrheliometers measure **global horizontal irradiance (GHI)** and **direct normal irradiance (DNI)** at specific locations.

2. Statistical and Empirical Models

- **Liu and Jordan Model**: Estimates hourly and daily solar radiation using clearness index.
- **Angstrom-Prescott Model**: Relates sunshine duration to solar radiation for site-specific estimation.

3. Numerical Weather Prediction (NWP) Models

Uses weather forecasting techniques (e.g., **WRF-Solar** developed by NOAA) to predict solar irradiance under varying atmospheric conditions.

5.6.1.2 Assessing Wind Speed and Power Potential

Wind energy resource assessment requires analyzing wind speed, direction, and turbulence intensity.

1. Wind Measurement Techniques

- **Meteorological Towers (Met Masts)**: Installed at heights of 60–120 m to collect long-term wind speed data.
- **LIDAR and SODAR Technology**: Remote sensing tools that measure wind profiles up to 200 m, essential for offshore and high-altitude assessments.

2. Wind Speed Distribution Models

- **Weibull Distribution**: The most common statistical model for wind resource assessment, defining wind speed probability at a site.
- **Rayleigh Distribution**: A simplified version of Weibull, assuming a shape factor of 2.

3. Wind Power Density Calculation

Wind power potential is estimated using:

$$P = \frac{1}{2}\rho A V^3 \quad \text{where} P \text{ is power, } \rho \text{ is air density, } A \text{ is the rotor-swept area,}$$

and V is wind speed.

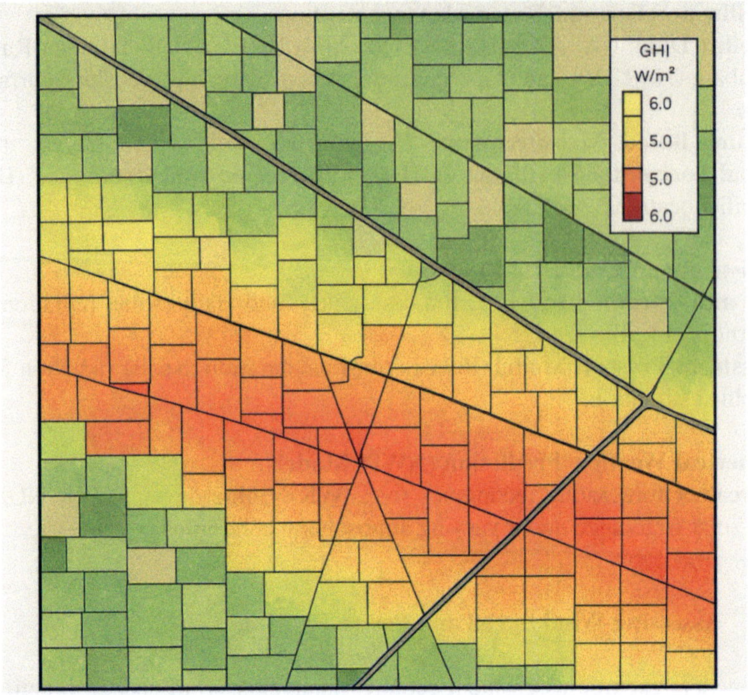

Fig. 5.19 Overlay Solar Radiation with Land Availability

5.6.2 Resource Mapping and GIS

Geographic Information Systems (GIS) play a crucial role in renewable energy site selection by integrating multiple geospatial datasets, including solar irradiation, wind potential, land use, grid proximity, and environmental constraints.

1. **GIS-Based Solar and Wind Resource Mapping**
   - **Solar Resource Maps**: Utilize datasets such as **NSRDB**, **PVGIS**, and **Copernicus** to overlay solar radiation with land availability (Fig. 5.19).
   - **Wind Resource Maps**: Derived from **Global Wind Atlas**, **NREL Wind Toolkit**, and **ERA5**, providing wind speed and power density distributions.
2. **Multi-Criteria Decision Analysis (MCDA) in GIS**
   - **Analytical Hierarchy Process (AHP)**: Weighs multiple factors, including solar radiation, slope, land use, and grid proximity.
   - **Weighted Overlay Analysis**: Assigns scores to geographic layers, generating suitability maps for wind and solar farms.

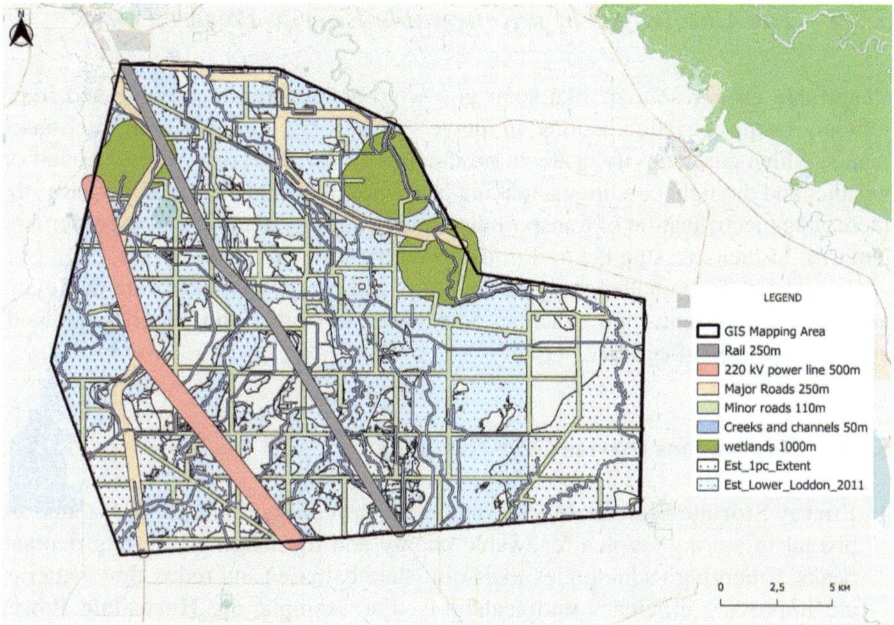

Fig. 5.20 Constraints Map from Willatook Wind Farm in Australia [36]

3. **Constraints and Exclusion Criteria**
   - **Environmental Regulations**: Protected habitats, water bodies, wetlands, etc.
   - **Land Use Restrictions**: Urban zones, agricultural lands, culturally sensitive and scenic areas.
   - **Infrastructure Proximity**: Distance to transmission lines, substations, residential structures, and roads (Fig. 5.20).

5.7 Grid Flexibility and Demand Response

As renewable energy continues to penetrate global energy markets, achieving grid flexibility and enhancing demand response mechanisms become paramount. Grid flexibility refers to the ability of the power system to manage variability and uncertainty in generation and load, especially under a high penetration of renewable energy. Demand response (DR) mechanisms empower consumers to adjust their electricity usage during peak demand or in response to grid constraints. Together, these innovations enable a more resilient, efficient, and sustainable energy grid.

5.7.1 The Challenges of a Renewable Energy-Powered Grid

Renewable energy sources like solar and wind are inherently variable and intermittent, leading to fluctuations in power generation. Unlike fossil fuel-based plants, which can ramp up or down production on demand, renewables depend on weather and daylight conditions, making grid stability a challenge. Additionally, the increasing electrification of transportation and heating systems creates new demand patterns, further stressing the traditional grid infrastructure [16].

To address these challenges, grid operators and energy innovators are deploying advanced technologies and strategies to enhance flexibility and integrate demand-side participation effectively.

5.7.1.1 Innovations Driving Grid Flexibility

1. **Energy Storage Systems (ESS):** Batteries, particularly lithium-ion systems, are pivotal in storing surplus renewable energy and discharging it during demand peaks. Emerging technologies like solid-state batteries and redox flow batteries are improving efficiency and scalability. For example, the **Hornsdale Power Reserve** in South Australia, equipped with Tesla's battery technology, has demonstrated the ability to stabilize the grid and reduce costs significantly [37].
2. **Advanced Grid Management Technologies:**
    - **Artificial Intelligence (AI) and Machine Learning (ML):** Predictive algorithms analyze weather patterns and demand forecasts, optimizing the dispatch of renewable energy and storage resources [10].
    - **Dynamic Line Rating (DLR):** Sensors monitor transmission line capacity in real-time, allowing grid operators to maximize the usage of existing infrastructure without overloading it.
    - **Microgrids:** Decentralized systems can operate independently or connect to the main grid, providing localized resilience and efficient renewable energy utilization.
3. **Flexible Generation Assets:** Modern natural gas turbines and hybrid power plants (combining renewables with storage or gas) offer quick ramping capabilities, ensuring grid stability during renewable energy lulls.

5.7.2 Demand Response Mechanisms

Demand response programs aim to align consumer electricity usage with grid conditions, reducing the need for peaking power plants and supporting renewable integration. Innovations in this area include:

1. **Smart Metering and Real-Time Pricing:** Advanced metering infrastructure (AMI) provides consumers with real-time data on electricity usage and costs. Time-of-use (TOU) pricing incentivizes shifting energy use to off-peak hours, flattening demand curves [38].
2. **Automated Demand Response (ADR):** Smart appliances, thermostats, and industrial control systems can automatically adjust energy consumption based on grid signals. For example, HVAC systems in large buildings may pre-cool spaces before peak periods or reduce cooling during demand spikes.
3. **Vehicle-to-Grid (V2G) Integration:** Electric vehicles (EVs) can serve as mobile energy storage units, discharging power back to the grid during high demand and charging when renewable generation is abundant [39].
4. **Behavioral Demand Response:** Gamification and personalized energy-saving recommendations encourage consumers to voluntarily reduce usage during peak periods. Digital platforms and mobile apps enhance participation through user-friendly interfaces and rewards.

5.7.3 Case Studies: Real-World Applications

1. **California ISO (CAISO):**
   The California Independent System Operator (CAISO) has become a global leader in managing grid flexibility. Faced with a significant penetration of solar generation, CAISO developed the Flexible Ramping Product, a market-based mechanism that ensures sufficient capacity is available to meet sudden changes in demand or supply. Additionally, CAISO's demand response programs enable commercial and residential customers to reduce their energy use during peak times. The "Duck Curve" phenomenon, characterized by steep ramping requirements during sunset, highlights the importance of integrating demand-side resources effectively [40]
2. **Energy Web Foundation (EWF):**
   EWF utilizes blockchain technology to create decentralized demand response markets. By leveraging this innovation, consumers and businesses can participate in peer-to-peer energy trading and contribute to grid flexibility. In a pilot project in Austria, EWF demonstrated how blockchain-enabled platforms could manage distributed energy resources, such as rooftop solar systems, in real-time [39].
3. **Denmark's Wind Integration:**
   Denmark's grid stands as a model for integrating high shares of wind energy, exceeding 50% in its energy mix. Through demand response programs and interconnectivity with neighboring countries, Denmark balances its renewable energy generation. The country uses real-time pricing to encourage industries and households to shift energy usage to periods of high wind power generation. Additionally, district heating systems are integrated with the grid to absorb surplus electricity [41].

Building on these innovations in grid flexibility and demand-side management, the next phase in the renewable energy transition involves harnessing advanced digital infrastructure to optimize grid performance. Smart grid technologies are now at the forefront, enabling real-time monitoring, two-way communication, and intelligent control across increasingly complex power systems. These systems allow for seamless integration of intermittent renewable sources, improved demand response, and more precise energy dispatch. The following section delves into how smart grid integration and analytics-driven dispatch optimization are enabling a more adaptive, resilient, and efficient energy grid.

5.8 Grid Integration and Dispatch Optimization

5.8.1 Smart Grid Technology and Renewables

With the increasing adoption of renewable energy sources, the need for advanced technologies that facilitate the seamless integration of these intermittent resources into the power grid has become more apparent. Smart grid technology is a key enabler of this transition, offering capabilities that enhance grid reliability, efficiency, and sustainability.

A smart grid is an upgraded electrical grid that uses digital communication technology to monitor and control the flow of electricity from all generation sources to meet the varying electricity demands of end users. This technology supports two- way communication between utilities and consumers, as well as real-time monitoring and control of electricity distribution. The integration of renewable energy sources, such as solar and wind, is particularly challenging due to their inherent variability and uncertainty. Smart grids use a combination of advanced sensors, automated control systems, and AI to address these challenges effectively [42].

Key features of smart grid technology include:

1. Advanced Metering Infrastructure (AMI): Enables real-time data collection on energy consumption and generation, allowing for better load forecasting and management.

2. Demand Response (DR): Facilitates dynamic changes in electricity usage in response to supply conditions, helping to balance supply and demand.

3. Distributed Energy Resources (DERs) Integration: Supports decentralized energy generation, such as solar panels and small wind turbines, improving grid resilience and reducing transmission losses.

4. Grid Management Systems: Enhance the visibility and control of grid operations, optimizing the dispatch of renewable energy and minimizing the impact of fluctuations in renewable generation [43].

By utilizing smart grid systems, utilities can better match renewable energy supply with consumer demand, mitigate the effects of intermittency, and improve grid stability. This ensures that renewable energy can be efficiently integrated into the grid without compromising reliability [42].

5.8.2 Demand Response and Load Balancing

The integration of renewable energy into the grid introduces new challenges due to its variable and unpredictable nature. Demand response (DR) is a crucial strategy to address these challenges. DR programs encourage consumers to adjust their energy consumption patterns in response to signals from the utility, such as price fluctuations or system constraints. This helps balance supply and demand in real-time, reducing the need for expensive peaking power plants and preventing grid congestion [44].

Techniques for implementing demand response include:

1. Time-of-Use Pricing: Encourages consumers to use electricity during off-peak times when renewable generation is abundant, thus helping to flatten peak demand periods.

2. Direct Load Control: Allows utilities to remotely control appliances, such as air conditioners or water heaters, to reduce energy consumption during high-demand periods.

3. Incentive-Based Programs: Consumers are financially incentivized to reduce or shift their energy usage during critical periods when grid stability is at risk.

4. Automated Demand Response: Leveraging IoT-enabled devices to automatically adjust energy consumption based on real-time grid conditions, improving overall system flexibility [45].

In addition to demand response, load balancing is essential for maintaining grid stability with high renewable energy penetration. Load balancing involves continu- ously adjusting the supply of electricity to match the demand. This requires advanced forecasting tools and predictive algorithms to anticipate energy demand and supply fluctuations.

Key methods for load balancing include:

- Forecasting renewable energy generation: Predicting the available output from wind and solar power, using weather data and historical trends to inform grid operators about expected generation levels.
- Flexible generation assets: Dispatching gas or hydroelectric plants that can quickly ramp up or down to balance renewable generation.
- Interconnection with neighboring grids: Leveraging regional grid interconnections to import or export electricity during periods of surplus or scarcity, ensuring a stable supply [46].

By optimizing both demand response and load balancing, utilities can mitigate the variability of renewable energy sources and ensure that grid stability is maintained. Storage and Grid Stability Energy storage systems play a pivotal role in enhancing grid stability, particularly with the high penetration of variable renewable energy sources like wind and solar.

These sources are highly intermittent, meaning that their energy output fluctuates based on weather patterns and time of day. Energy storage technologies, such as batteries, pumped hydro storage, and compressed air energy storage (CAES), provide a buffer that allows excess energy to be stored during periods of high renewable generation and dispatched when generation is low or demand spikes [47].

The integration of storage into the grid offers several advantages:

1. Load leveling: Storage systems help smooth out fluctuations in renewable energy supply by storing excess energy during peak production and releasing it during low production or high demand.
2. Frequency regulation: Storage systems can quickly respond to fluctuations in grid frequency, providing vital services such as primary and secondary frequency control, ensuring the grid operates within its required frequency range [48].
3. Peak shaving: Storage systems can discharge electricity during peak demand periods to reduce the need for additional generation from fossil-fuel plants, thus reducing emissions and increasing the overall sustainability of the grid. Additionally, pumped hydro storage (PHS) and batteries offer grid operators the ability to respond to frequency imbalances in milliseconds, which is particularly valuable with high renewable penetration. With a rapid response time, energy stor- age systems can smooth out the impacts of short-term fluctuations in renewable generation, contributing to grid reliability [24].

5.8.3 Case Studies: Real-World Applications of Analytics

In the rapidly evolving field of renewable energy, analytics plays a pivotal role in enhancing the efficiency, reliability, and performance of various energy systems. The following case studies highlight how companies are leveraging advanced data analytics to optimize their renewable energy projects, including solar, wind, and hybrid systems.

5.8.3.1 Solar Energy Optimization Using AI and Data Analytics

Nellis Solar Power Plant (NV Energy, First Solar)

The Nellis Solar Power Plant in Nevada, USA, is one of the largest and most advanced solar power facilities in the United States. Managed by NV Energy, a subsidiary of NextEra Energy, the project has benefited from the integration of advanced analytics for performance optimization. First Solar, the primary contractor for the plant, used predictive analytics to optimize the solar panels' performance, focusing on maximizing energy production while minimizing maintenance costs. Data analytics tools were used to monitor weather patterns, predict solar irradiance, and adjust the operations of the panels to ensure maximum efficiency throughout the day.

The incorporation of machine learning algorithms allowed for the analysis of historical performance data, enabling the plant to optimize energy dispatch schedules based on real-time forecasts of energy demand and supply fluctuations [49]. This case study demonstrates how big data can significantly improve the operational efficiency of large-scale solar projects.

5.8.3.2 Wind Energy Dispatch Optimization

Hornsea Project One (Orsted)

The Hornsea Project One, developed by Orsted, is one of the largest offshore wind farms globally, located off the coast of the United Kingdom. This wind project utilized advanced analytics for grid integration and dispatch optimization. Orsted employed predictive analytics and real-time data monitoring to forecast wind energy production and adjust the operation of wind turbines accordingly [50].

By using machine learning models to analyze meteorological data, the company was able to optimize turbine operations, improve grid stability, and reduce the costs associated with overproduction or underproduction of energy. Through dynamic scheduling based on wind forecasts, the project demonstrated how analytics could help wind farms maximize energy output while minimizing waste [50]. This case exemplifies the use of advanced data analytics in the wind energy sector to improve the reliability and efficiency of grid integration, ensuring that renewable energy generation is maximized without overwhelming the grid.

5.8.3.3 Analytics-Driven Hybrid Renewable Systems

Juno Beach Hybrid System (Florida Power and Light, NextEra Energy)

The Juno Beach Hybrid System developed by Florida Power & Light (FPL), a subsidiary of NextEra Energy, combines solar, wind, and energy storage technologies to create a reliable and efficient hybrid renewable energy system. The project is located in Florida, and it integrates energy from multiple sources to ensure that the grid remains stable and reliable. FPL leveraged analytics to monitor and optimize the performance of solar and wind energy generation in real-time, and also to manage energy storage systems that can store excess energy for later use. Predictive analytics and optimization algorithms were used to forecast energy demand and adjust the energy dispatch from various sources accordingly.

The hybrid system's performance was continuously analyzed using data from advanced sensors and control systems to optimize the energy mix and reduce reliance on fossil fuels [51].

Through the use of data analytics, FPL has been able to create a robust energy system that not only reduces carbon emissions but also improves the reliability and economic efficiency of energy supply. The hybrid model demonstrates the potential for integrating multiple renewable energy sources and energy storage to create a more resilient and sustainable grid.

The Path Forward

The transition to a renewable energy-powered grid requires not just technological advancements but also regulatory and market reforms. Policymakers must:

- Encourage investments in energy storage, smart grid technologies, and DR programs.
- Develop standardized frameworks for DER and V2G integration.
- Facilitate cross-border energy trading to leverage regional resource diversity.

Simultaneously, public awareness and participation are critical. Empowering consumers with tools and incentives to actively engage in demand response will bridge the gap between renewable generation and consumption.

References

1. MBT Energy, Floating photovoltaic power generation: a new direction for solar energy. https://www.mbt-energy.com/news/industry/2207151.html (2022). Accessed 03 May 2025
2. Sara Wolf, What are bifacial solar panels? https://www.paradisesolarenergy.com/blog/what-are-bifacial-solar-panels (2020). Accessed 03 May 2022.
3. H.H. Park, Transparent electrode techniques for semitransparent and tandem perovskite solar cells. Electron. Mater. Lett. **17**, 18–32 (2021)
4. Oxford PV, Oxford pv sets new solar cell world record (2023). Accessed 04 May 2025
5. ESS Solar Power, Thin film solar panels: How they work and applications (2023). Accessed 04 May 2025
6. C. Prieto, A. Blindu, L.F. Cabeza, J. Valverde, G. García, Molten salts tanks thermal energy storage: aspects to consider during design. Energies **17**(1), 22 (2024)
7. E. Hau, *Wind Turbines: Fundamentals, Technologies, Application, Economics*, 3rd edn. (Springer, Berlin, 2013)
8. U.S. Department of Energy, Wind turbines: Bigger and better (2022). Accessed 04 May 2025
9. P. Veers et al., Grand challenges in the science of wind energy. Science **366**, 1–8 (2019)
10. International Energy Agency, World energy outlook 2021. https://www.iea.org (2021)
11. J. Jonkman et al., Design and analysis of wind turbine rotor blades. National Renewable Energy Laboratory (NREL) (2018)
12. Wikimedia Commons contributors, Uk's largest onshore turbine blades (2020). Accessed 04 May 2025
13. W. Musial, P. Beiter, P. Duffy, Floating offshore wind energy: technical advancements and market growth. National Renewable Energy Laboratory (2022)
14. A. Myhr, C. Bjerkseter, A. Ågotnes, Cost reduction pathways for floating offshore wind turbines. Energy Policy **134**, 110–123 (2021)
15. M. Borg, R. Patel, H. Jensen, Artificial intelligence in offshore wind energy systems. Renew. Energy J. **95**(3), 123–139 (2023)
16. IRENA, Innovation Outlook: Offshore Wind. International Renewable Energy Agency (2022)
17. Transformers Magazine, China completes world's first hybrid offshore wind-solar power plant (2023). Accessed 04 May 2025
18. Sonardyne, Why are your dynamic cables behaving like this? (2023). Accessed 04 May 2025
19. International Energy Agency (IEA), *Offshore Wind Outlook 2023* (IEA, Paris, 2023)

20. WindEurope, *Floating Wind: Opportunities for Europe's Energy Transition* (WindEurope Publications, Brussels, 2023)
21. P. Bortolotti et al., Aerodynamic Optimization of Large Wind Turbines. Wind Energy Science (2021)
22. H. Chen et al., Progress in electrical energy storage system: a critical review. Prog. Nat. Sci. **19**(2), 291–303 (2009)
23. L.Zhang et al., Advances in gravity-based energy storage technologies. Renew. Sustain. Energy Rev. **155**, 111877 (2023)
24. T.Liu et al., Recent developments in vanadium redox flow batteries for large-scale energy storage applications. J. Power Sources **451**, 227821 (2020)
25. Y. Zhou et al., Vanadium redox flow battery for energy storage: a review. J. Energy Storage **24**, 100786 (2019)
26. M. Bolinger, J. Seel, D. Millstein, Utility-scale solar, 2023 edition. Technical report, Lawrence Berkeley National Laboratory (2023)
27. California Energy Commission, Levelized cost of electricity and other metrics for renewable energy technologies (2022)
28. Plug and Play Tech Center, Harnessing earth's heat: Applications of geothermal energy (2024) Accessed 05 May 2025
29. U.S. Energy Information Administration, Where geothermal energy is found (2024). Accessed 05 May 2025
30. J.W. Lund, South dakota geothermal resources. Geo-Heat Center Q. Bull. **18**(4), 1–6 (1997)
31. U.S. Energy Information Administration, Geothermal power plants (2024). Accessed 05 May 2025
32. T. Navidi, A. El Gamal, R. Rajagopal, Coordinating distributed energy resources for reliability can significantly reduce future distribution grid upgrades and peak load. Joule **7**(8), 1769–1792 (2023)
33. M. Amin, Demand Response in Smart Grids. MDPI (2021)
34. R. Kumar, H. Singh, iot-based monitoring and predictive maintenance of renewable energy systems: a review. Renew. Sustain. Energy Rev. **153**, 111758 (2022)
35. National Renewable Energy Laboratory, What is the nsrdb? (2024). Accessed 11 May 2025
36. GIS-Point, Analysis for determining optimal locations (n.d.). Accessed 11 May 2025
37. Australian Energy Market Operator, Hornsdale power reserve impact assessment. https://aemo.com.au (2021). Retrieved from AEMO
38. U.S. Department of Energy, Wind energy and the environment: addressing impacts and mitigation strategies. https://www.energy.gov/eere/wind/wind-energy-environment (2020)
39. Energy Web Foundation, Blockchain for grid flexibility. https://energyweb.org (2022). Retrieved from EWF
40. California Independent System Operator, Flexibility for a renewable grid. https://caiso.com (2021). Retrieved from CAISO
41. International Energy Agency, Denmark 2023: Energy policy review (2023)
42. H. Liu, L. Liu, X. Liu, Integration of renewable energy sources into smart grids. Renew. Energy **137**, 623–634 (2019)
43. P. Siano, Demand response and smart grids—a survey. Renew. Sustain. Energy Rev. **30**, 461–478 (2014)
44. P. Palensky, D. Dietrich, Demand side management: demand response, intelligent energy systems, and smart loads. Proc. IEEE **99**(1), 144–152 (2011)
45. S. Hopper, M. Dailey, A. Szydlowski, Demand response for renewable energy integration: a review of challenges and opportunities. Renew. Sustain. Energy Rev. **55**, 61–69 (2016)
46. P. Denholm, M. Hand, Grid flexibility and storage required to achieve high penetration of variable renewable electricity. Energy Policy **39**(3), 1813–1820 (2011)
47. Q. Zhou, X. Ma, J. Zhao, Advanced energy storage technologies and their role in modern electric grids. Energy Rep. **6**, 764–779 (2020)
48. Y. Zhang et al., Advances in Solid-State Electrolytes for Lithium Batteries. Nature Reviews Materials, 2021.

49. A. Gonzalez et al., Optimization and performance enhancement of the Nellis solar power plant using data analytics. Renew. Energy J. **24**(3), 157–163 (2019)
50. Orsted, Hornsea project one: Data-driven wind farm performance optimization. https://orsted.com (2020). Retrieved from Orsted
51. Florida Power and Light, Juno Beach Hybrid System: Leveraging Analytics for Renewable Energy Optimization. NextEra Energy Insights (2021)

Chapter 6
Financial Modeling and Economic Analysis

6.1 Financial Metrics for Renewable Projects

In the rapidly evolving renewable energy industry, financial metrics play a vital role in determining the viability and attractiveness of projects. Investors, developers, and policy makers rely on these metrics to assess risks, predict returns, and make informed decisions. This chapter delves into three key financial metrics: Internal Rate of Return (IRR), Net Present Value (NPV), and Levelized Cost of Energy (LCOE). These metrics offer a comprehensive perspective on profitability, investment potential, and cost efficiency in renewable energy projects. The chapter also explores methods for assessing the feasibility of renewable energy projects, identifies associated risks and sensitivity factors, and outlines various strategies for financing the projects.

6.1.1 Internal Rate of Return (IRR)

The Internal Rate of Return (IRR) represents the discount rate that equalizes the net present value (NPV) of cash inflows and outflows of a project to zero. Essentially, IRR is the annualized rate of return a project is expected to generate. A higher IRR indicates a more attractive project for investors, as it reflects the potential for higher returns relative to the cost of capital.

D. Gupta et al., *The Roadmap to Renewable Energy*,
https://doi.org/10.1007/978-3-032-09609-8_6

6.1.2 *Calculation of IRR*

The IRR is calculated by solving the following equation for r (the discount rate):

$$\sum_{t=0}^{n} \frac{C_t}{(1+r)^t} = 0$$

Where:

- C_t: Cash flow at time t
- r: Internal Rate of Return
- n: Total number of periods

Modern tools like Excel, financial modeling software, and programming languages (e.g., Python or R) simplify IRR calculation.

6.1.3 *Illustrative Example: Onshore Wind Project*

Consider a 50 MW onshore wind project that requires an upfront investment of $50 million. Over 20 years, it is expected to generate annual revenue of $7 million and incur $1 million in operational expenses. Using financial modeling, the IRR is calculated to be 10.5%. If the project's cost of capital is 8%, this IRR suggests that the project is financially viable, yielding a return higher than its financing costs.

6.1.4 *Limitations and Considerations of IRR*

- **Uncertainty in Cash Flows:** Renewable projects often face variability in revenues due to weather-dependent energy generation.
- **Sensitivity Analysis:** Developers perform sensitivity analyses, adjusting assumptions for variables like energy prices and operational costs.

6.1.5 *Net Present Value (NPV)*

NPV calculates the net value added by a project, taking into account the time value of money. Unlike IRR, which focuses on percentage returns, NPV provides a dollar figure that directly measures profitability.

6.1.6 Formula for NPV

$$NPV = \sum_{t=0}^{n} \frac{C_t}{(1+r)^t} - C_0$$

Where:

- C_t: Cash flow at time t
- r: Discount rate (cost of capital)
- C_0: Initial investment

A positive NPV indicates that a project is expected to generate value above its cost of capital, while a negative NPV suggests the opposite.

6.1.7 Case Study: Solar PV Project

A 20 MW solar PV farm requires an upfront cost of $20 million and is expected to generate $3 million annually for 25 years. Using a 6% discount rate, the NPV is calculated as $10 million. This positive NPV demonstrates that the project not only covers its cost of capital but also generates significant additional value, making it an attractive investment.

6.1.8 Role of NPV in Investment Decision-Making

- **Securing Financing:** Banks and investors often require positive NPVs as a prerequisite for lending.
- **Portfolio Optimization:** Developers compare the NPVs of multiple projects to allocate resources efficiently.

6.1.9 Levelized Cost of Energy (LCOE)

The Levelized Cost of Energy (LCOE) is the average cost of generating one unit of electricity (e.g., per kilowatt-hour) over the lifetime of a project. It incorporates all expenses, including capital costs, operational costs, and maintenance costs, normalized over the energy produced.

Table 6.1 presents indicative levelized cost of energy (LCOE) ranges for selected renewable energy technologies based on widely cited industry benchmarks.

Table 6.1 Comparison of LCOE across renewable energy technologies (based on Lazard [1])

Technology	Average LCOE ($/MWh)	Key drivers
Solar PV	30–50	Declining module costs, efficiency improvements
Onshore Wind	25–40	High capacity factors, streamlined turbine designs
Geothermal	40–60	Exploration risks, high upfront investment

6.1.10 LCOE Formula

$$LCOE = \frac{\sum_{t=0}^{n} \frac{C_t}{(1+r)^t}}{\sum_{t=0}^{n} \frac{E_t}{(1+r)^t}}$$

Where:

- C_t: Costs at time t
- E_t: Energy produced at time t
- r: Discount rate

6.1.11 Comparative Analysis of IRR, NPV and LCOE

IRR, NPV, and LCOE each evaluate renewable energy projects from different financial perspectives and should be interpreted in combination. IRR is useful for comparing percentage returns but does not reflect absolute value creation and can be misleading when project scales differ. NPV addresses this limitation by measuring total value added in monetary terms, making it critical for investment and portfolio decisions, though it is sensitive to discount rate assumptions. LCOE provides a standardized cost benchmark for comparing technologies and informing policy discussions, but it does not capture revenue structures, market exposure, or project-specific risks. Consequently, no single metric is sufficient on its own, and a comprehensive financial assessment requires the integrated use of all three metrics.

6.1.12 Role of LCOE in Policy and Market Design

Governments often use LCOE as a benchmark to allocate subsidies and determine feed-in tariffs. For instance, solar and wind technologies have seen rapid adoption due to their declining LCOEs and favorable policy frameworks.

IRR, NPV, and LCOE each offer distinct yet complementary insights into the financial performance of renewable energy projects. By leveraging these metrics, developers and investors can navigate the complexities of renewable energy finance, ensuring sustainable growth and competitive advantage in an ever-expanding industry.

6.2 Economic Feasibility of Renewable Projects

The economic feasibility of renewable energy projects is a critical factor in determining their viability and appeal to investors, developers, and policy-makers. Renewable projects often require substantial initial investments, but their long-term benefits, including low operational costs and environmental sustainability, make them attractive. Below, we delve deeper into the key aspects of economic feasibility: **CapEx and OpEx, Break-Even Analysis**, and **Payback Period**.

6.2.1 Capital Expenditure (CapEx) and Operational Expenditure (OpEx)

1. **Capital Expenditure (CapEx):**
   - **Definition:** CapEx encompasses the upfront costs associated with setting up a renewable energy project. These costs often represent a large portion of the total investment.
   - **Components:**
     - **Land Acquisition:** Costs for purchasing or leasing land, which vary based on location and availability. For instance, remote locations may have lower land costs but higher logistical expenses.
     - **Equipment Procurement:** This includes purchasing key components like wind turbines, solar panels, inverters, and mounting systems.
     - **Construction and Installation:** Costs for building infrastructure, installing equipment, and connecting to the grid.
     - **Permitting and Licensing:** Securing legal permissions and meeting regulatory requirements often incurs significant expenses, especially in regions with complex permitting processes.
2. **Operational Expenditure (OpEx):**
   - **Definition:** OpEx refers to the ongoing costs of operating and maintaining a project over its lifetime. While OpEx is generally lower than CapEx for many renewable technologies, it remains a critical factor in economic feasibility.
   - **Examples of OpEx:**
     - Routine maintenance of turbines or panels.
     - Staff salaries and training for operation and monitoring.
     - Insurance to protect against damage or liability.
     - Compliance costs, such as meeting safety and environmental standards.

Impact on Financial Viability

- Projects with **high CapEx and low OpEx** (e.g., solar PV) become profitable over time due to minimal operational costs.

- Conversely, projects with **moderate CapEx but high OpEx** (e.g., offshore wind) require rigorous financial planning and favorable market conditions to remain sustainable.
- Effective CapEx and OpEx management are essential for reducing financial risks and ensuring long-term profitability.

6.2.2 Break-Even Analysis

Understanding the Break-Even Point

- **Definition:** The break-even point is the moment when cumulative revenues from the project equal its total costs, including CapEx and OpEx. Beyond this point, the project starts generating net profits.
- **Importance:** It provides insight into the financial health of a project and helps stakeholders evaluate the timeline required to achieve profitability.

Key Determinants of Break-Even

- **Energy Production Rates:** The higher the energy output, the faster the revenues accumulate, shortening the break-even period. For instance, wind farms in high-wind areas break even quicker than those in moderate-wind locations.
- **Market Conditions:** Power Purchase Agreements (PPAs) and competitive electricity rates directly affect revenue streams.
- **Operational Efficiency:** Efficient resource management and lower OpEx accelerate the break-even process.

Case Examples

- **Solar PV:** A utility-scale solar project in a sunny region may break even in 6–8 years, depending on installation costs and market prices.
- **Offshore Wind:** Offshore wind farms often require 10–12 years to break even due to higher upfront costs and maintenance expenses.

6.2.3 Payback Period Analysis

Significance of the Payback Period

- The payback period measures how long it takes for a project to recover its initial investment. It is a crucial metric for investors as it highlights financial risk and return on investment.
- Shorter payback periods are generally preferred as they reduce exposure to uncertainties such as fluctuating energy prices or policy changes.

Factors Influencing the Payback Period

1. **Technology:**
   - **Solar PV:** With declining costs of panels and simpler installations, solar projects often have shorter payback periods compared to other technologies.
   - **Geothermal:** Exploration and drilling costs for geothermal energy can lead to longer payback periods, but the stable output offers long-term returns.
   - **Wind:** Onshore wind projects typically have moderate payback periods due to relatively balanced CapEx and OpEx.
2. **Location:**
   - Projects in regions with abundant renewable resources (e.g., high solar insolation or strong wind) achieve faster payback periods.
   - Accessibility to infrastructure and proximity to energy markets also play a role.
3. **Government Support and Policies:** Subsidies, tax incentives, and favorable policies have historically played a crucial role in reducing the payback period of renewable energy projects. For example, Production Tax Credits (PTC) for wind energy in the U.S. have significantly boosted project feasibility. However, these incentives currently face uncertainty and potential reduction under changing political and administrative priorities, which could slow the momentum of renewable energy development unless new, stable policy frameworks are established.
4. **Financing Structure:** Lower interest rates and innovative financing mechanisms, such as green bonds or public-private partnerships, can enhance economic feasibility by reducing financial burdens.

Synthesis
Economic feasibility is central to the successful deployment of renewable energy projects. By understanding and optimizing CapEx and OpEx, accurately analyzing break-even points, and minimizing payback periods, developers and investors can ensure that projects not only meet sustainability goals but also deliver robust financial returns. These metrics also enable stakeholders to make informed decisions, fostering confidence in the transition to renewable energy solutions.

6.3 Risk and Sensitivity Analysis

Renewable energy projects operate in dynamic environments influenced by market fluctuations, regulatory changes, technological advancements, and operational risks. To mitigate uncertainties, developers and investors use **risk and sensitivity analysis** techniques. This chapter explores **sensitivity analysis in financial modeling, financial risk assessment, and scenario planning**, providing insights into how project developers evaluate and manage risks effectively.

6.3.1 Sensitivity Analysis in Financial Modeling

Sensitivity analysis is a quantitative technique used to determine how different values of input variables impact a given output variable in financial models. It helps project developers assess **financial viability under varying conditions** and identify critical variables that significantly affect project outcomes.[2]

6.3.1.1 Application of Sensitivity Analysis

In renewable energy project finance, sensitivity analysis is commonly used to evaluate how fluctuations in key parameters, such as **electricity prices, capital costs, interest rates, and capacity factors**, influence project returns. [3]

Key Steps in Sensitivity Analysis

1. **Identify Key Variables:** Determine the financial parameters that significantly impact cash flows and profitability.
2. **Define Input Ranges:** Establish realistic high and low estimates for each variable based on historical data and market trends.
3. **Run the Model:** Adjust one variable at a time while keeping others constant to observe the impact on key financial metrics such as **Net Present Value (NPV), Internal Rate of Return (IRR), and Levelized Cost of Energy (LCOE)**.
4. **Interpret Results:** Identify which variables introduce the most uncertainty and develop strategies to mitigate associated risks.

Example

A solar project's **NPV** is tested against varying electricity prices ($40/MWh to $80/MWh). Results show that if prices fall below $50/MWh, the project becomes unprofitable, highlighting **electricity price risk** as a key financial concern. [1]

6.3.1.2 Identifying Key Variables That Affect Project Outcomes

Sensitivity analysis helps identify the most influential factors affecting project success. Common key variables in renewable energy projects include:

- **Electricity Prices:** Revenue depends on market prices and long-term **Power Purchase Agreements (PPAs)**[1] [4]
- **Capital Expenditures (CapEx):** High upfront costs make renewable projects sensitive to cost overruns
- **Operating Expenses (OpEx):** Maintenance costs can fluctuate, impacting long-term financial sustainability.

[1] A Power Purchase Agreement (PPA) is a long-term contract between an energy producer and buyer to purchase electricity at a pre-agreed price.

- **Capacity Factor:** The actual energy output relative to maximum potential output.
- **Discount Rate:** Affects NPV calculations and overall investment attractiveness.

6.3.2 Financial Risk Assessment

6.3.2.1 Types of Financial Risks in Renewable Energy

Renewable energy projects face multiple financial risks, categorized into **market risks, operational risks, regulatory risks, and financial risks** [5].

Market Risk

- **Electricity Price Volatility:** Fluctuations in wholesale electricity markets affect revenue streams, especially for merchant projects.
- **Inflation and Interest Rates:** Rising interest rates increase debt servicing costs, making financing more expensive [6].

Operational Risk

- **Equipment Failures:** Unplanned downtime due to **wind turbine failures or solar panel degradation** leads to revenue losses [7].
- **O&M Cost Variability:** Higher-than-expected maintenance costs reduce project profitability [8].

Regulatory and Policy Risks

- **Changes in Subsidies and Incentives:** The expiration or reduction of **Investment Tax Credits (ITC)** and **Production Tax Credits (PTC)** impacts project feasibility [9].
- **Grid Connection Delays:** Permitting and interconnection bottlenecks can delay project timelines and increase costs [10].

Financial Risk

- **Credit Risk:** Off-takers defaulting on PPAs can lead to cash flow disruptions.
- **Liquidity Risk:** Difficulty in securing financing for project expansions or refinancing existing debt.

6.3.2.2 Strategies to Mitigate Financial Risks

To manage financial risks, developers and investors employ several mitigation strategies:

1. **Hedging[2] Against Market Risks:**
    - Entering long-term **PPAs** instead of relying on volatile spot markets [1].

---

[2] Hedging refers to securing long-term contracts or financial instruments to protect against fluctuations in energy prices and ensure cost stability.

- Utilizing **derivatives and hedging contracts** to protect against price fluctuations.

2. **Diversifying Revenue Streams:**

   - Combining revenue from **RECs (Renewable Energy Certificates), capacity markets, and ancillary services**.
   - Investing in **hybrid projects** (e.g., solar + storage) to improve grid flexibility.

3. **Robust Financial Modeling and Stress Testing:**

   - Running **Monte Carlo simulations** to assess financial impacts under uncertain conditions. [11]
   - Creating **worst-case and best-case financial projections** for investors.

4. **Leveraging Government Incentives:**

   - Structuring projects to maximize **ITC/PTC benefits** [12] (Kreycik, Couture, & Cory, 2011).
   - Applying for **low-interest loans and grants** to reduce financing costs.

6.3.3 Scenario Planning

6.3.3.1 Scenario-Based Financial Modeling

Scenario planning involves developing multiple financial projections based on different assumptions about market conditions, policy changes, and technological advancements. [13] This approach helps project developers prepare for **uncertainties and optimize decision-making**.

Types of Financial Scenarios

1. **Best-Case Scenario:** Assumes favorable conditions, such as high electricity prices, low interest rates, and strong government incentives.
2. **Base-Case Scenario:** Represents the most **realistic and likely** outcome based on current market trends.
3. **Worst-Case Scenario:** Considers unfavorable conditions, including high capital costs, low electricity prices, and policy uncertainty.

6.3.3.2 Example of Scenario Planning in a Wind Project

A **100 MW wind project** is modeled under three scenarios (Table 6.2):
This analysis helps **investors assess risk appetite** and developers adjust financial strategies accordingly [14].

Table 6.2 IRR and NPV of a 100MW wind project under different electricity price scenarios (author calculations)

Scenario	Electricity price ($/MWh)	IRR (%)	NPV ($M)
Best case	$70	12%	$50
Base case	$50	8%	$20
Worst case	$35	3%	-$10

6.3.4 Synthesis

Risk and sensitivity analysis are crucial for ensuring the financial stability of renewable energy projects. **Sensitivity analysis** helps identify key variables affecting financial outcomes, while **financial risk assessment** allows developers to recognize and mitigate market, operational, and regulatory risks. **Scenario planning** ensures that projects remain resilient under varying market conditions, enabling better decision-making for long-term success. By integrating these approaches into financial modeling, renewable energy developers can **optimize risk management, enhance investment confidence, and secure project sustainability**.

6.4 Financing Renewable Projects

Financing is one of the most critical aspects of renewable energy project development. The success of a project depends on securing adequate funding and structuring financial agreements that ensure long-term profitability. Various financial mechanisms, including **Power Purchase Agreements (PPAs), debt financing, equity structures, and tax incentives**, play a key role in making projects viable. This section explores these financing methods and how they impact the economic sustainability of renewable energy projects.

6.4.1 Power Purchase Agreements (PPAs)

A **Power Purchase Agreement (PPA)** is a long-term contract between an energy producer (the project owner) and an off-taker (a utility, corporation, or government entity) that defines the terms of electricity sales. PPAs are crucial for securing project financing, as they provide predictable revenue streams, reducing financial risk for investors and lenders [14].

6.4.1.1 Key Components of a PPA

1. **Contract Duration:** Typically ranges from **10 to 25 years**, providing long-term revenue certainty.
2. **Pricing Mechanism:** Can be **fixed price, indexed price**, or **escalating price**, depending on market conditions [4].
3. **Energy Delivery Terms:** Specifies the volume of electricity to be delivered and penalties for underperformance.
4. **Curtailment Provisions:** Defines compensation mechanisms if the off-taker limits energy intake due to grid constraints.

6.4.1.2 Long Term Contracts vs. Merchant Exposure

- **Long-Term PPAs:** Provide stability for both the developer and investor but may have lower profit potential compared to fluctuating market prices [15].
- **Merchant (Spot Market) Sales:** Allow developers to sell electricity at prevailing market prices, which can be riskier but potentially more profitable in high-demand periods [16].

6.4.1.3 Example

A **200 MW solar farm** signs a **20-year PPA** with a utility at a fixed rate of **$50/MWh**. This ensures a steady revenue stream, making it easier to secure loans. However, if market prices later rise to **$70/MWh**, the developer misses out on potential higher earnings due to the fixed contract [1].

6.4.2 Debt and Financing Structures

6.4.2.1 Debt Financing for Renewable Energy Projects

Debt financing involves borrowing funds to develop a project, typically through **bank loans, project bonds, or institutional lenders**. It is a common approach in renewable energy financing, as lenders are willing to invest in projects with stable cash flows [6].

Types of Debt Financing
1. **Commercial Bank Loans:** Traditional debt financing from banks, often secured by project assets [5].
2. **Green Bonds:** Debt instruments issued to finance environmentally sustainable projects.

3. **Government Loans:** Some governments offer **low-interest loans** or **loan guarantees** to promote renewable energy development [17].
4. **Project Finance (Non-Recourse Loans):** Debt secured against project revenues rather than developer balance sheets, reducing financial risk for the parent company.

6.4.2.2 Equity Financing for Renewable Projects

Equity financing involves raising capital by offering ownership stakes in the project. Investors, such as **venture capital firms, private equity funds, or institutional investors**, provide funding in exchange for future returns [18].

Types of Equity Investors

1. **Developers & Sponsors:** The project owner contributes equity to retain majority control.
2. **Institutional Investors:** Pension funds, insurance companies, and sovereign wealth funds seeking long-term returns.
3. **Strategic Investors:** Large energy companies investing to expand their renewable portfolio.

6.4.2.3 Optimal Capital Structure: Debt vs. Equity

Renewable energy developers must carefully balance debt and equity to optimize project returns while minimizing financial risk. A typical **capital structure** might involve **70% debt and 30% equity**, depending on project risk and investor requirements [5].

Example
A **100 MW wind project** requires **$200 million** in financing. A developer secures **$140 million (70%) in debt** and raises **$60 million (30%) in equity** from investors. This structure ensures lower capital costs while leveraging external funding for growth.

6.4.3 Tax Equity and Government Incentives

6.4.3.1 The Role of Tax Incentives in Renewable Energy Financing

Government incentives significantly impact the financial viability of renewable projects by **reducing upfront costs** and **attracting investors**. The two most common tax incentives in the U.S. are:

1. **Investment Tax Credit (ITC):** Provides a **federal tax credit** based on a percentage of the project's capital costs. Solar projects can receive **up to 30% ITC**, directly reducing tax liabilities [9].
2. **Production Tax Credit (PTC):** Offers a **per-kilowatt-hour (kWh) tax credit** for electricity generated from wind, biomass, and other renewables over the first **10 years** of operation [10].

6.4.3.2 Tax Equity Financing: Leveraging Government Incentives

Since renewable developers may not always have sufficient tax liabilities to utilize credits, **tax equity investors**, such as banks and corporations, invest in projects to benefit from tax incentives [12].

How Tax Equity Financing Works

1. A tax equity investor provides capital in exchange for the right to claim **ITC or PTC benefits**.
2. The developer retains operational control but shares a portion of cash flows with the tax equity investor.
3. Once tax benefits expire (typically after **5–7 years**), the investor exits the project, allowing the developer to regain full ownership.

6.4.3.3 Example

A **50 MW solar farm** costing **$100 million** qualifies for a **30% ITC ($30 million tax credit)**. A large bank invests **$30 million** in exchange for the tax benefits, reducing the developer's financial burden while securing a return on investment through tax savings.

6.4.3.4 Additional Government Incentives

- **Renewable Energy Certificates (RECs):** Tradable credits that allow utilities to meet renewable energy mandates.
- **State-Level Incentives:** Grants, property tax exemptions, and production subsidies vary by region [19] (DSIRE, 2023).
- **Direct Subsidies:** Some governments provide **cash rebates** or **low-interest loans** to renewable projects [17].

6.4.4 Synthesis

Securing financing is essential for the successful deployment of renewable energy projects. **Power Purchase Agreements (PPAs)** provide financial stability by securing long-term revenue, while **debt and equity financing** offer various capital-raising options tailored to project risk. Government incentives such as **ITC and PTC** play a crucial role in attracting tax equity investors, making projects more financially viable. By strategically combining these financing mechanisms, renewable energy developers can **mitigate risks, optimize capital structures, and ensure long-term profitability**.

6.5 Revenue Models for Renewable Projects

Renewable energy projects generate revenue through multiple mechanisms, including **merchant market sales, long-term contracts, capacity markets, ancillary services, and Renewable Energy Certificates (RECs)**. The chosen revenue model significantly impacts the financial stability and profitability of a project. This section explores these models in detail, assessing their risks and benefits.

6.5.1 Merchant vs. Contracted Revenue Streams

6.5.1.1 Merchant Market Revenue

In the **merchant market**, renewable energy producers sell electricity at prevailing market prices without long-term contracts. This exposes projects to price volatility but allows them to capture higher revenues during peak pricing periods [20]. Merchant projects rely on **wholesale electricity markets**, where pricing is influenced by supply-demand dynamics, regulatory changes, and fuel costs [21].

Advantages of Merchant Revenue

- Higher revenue potential in high-price environments.
- Flexibility to take advantage of market fluctuations.
- No contractual limitations on electricity sales.

Risks of Merchant Revenue

- Price volatility can result in revenue unpredictability.
- Difficult to secure financing due to uncertain cash flows [22].

6.5.1.2 Contracted Revenue (Power Purchase Agreements: PPAs)

Power Purchase Agreements (PPAs) provide a **stable revenue stream** by securing long-term electricity sales contracts, typically ranging from 10 to 25 years [10]. Utilities, corporations, or government entities often act as off-takers in PPAs.

Advantages of Contracted Revenue

- Predictable cash flows enhance bankability and project financing.
- Lower exposure to market price fluctuations.
- Attracts investors due to reduced financial risk.

Risks of Contracted Revenue

- Potentially lower revenue compared to volatile merchant prices.
- Limited ability to benefit from price spikes.
- PPA negotiations can be complex and lengthy.

6.5.2 Capacity Markets and Ancillary Services

6.5.2.1 Capacity Markets

Capacity markets compensate power generators for maintaining available capacity to meet future grid demand. These markets, primarily in regions like **PJM, NYISO, and ISO-NE**, provide payments to generators to ensure long-term reliability [23].

Revenue Potential from Capacity Markets

- Wind and solar projects can participate by offering firm capacity backed by storage.
- Grid operators procure capacity years in advance, providing revenue stability.

Challenges in Capacity Markets

- Intermittent renewable sources may require firming solutions, such as batteries or hybrid systems.
- Capacity accreditation rules can limit earnings for renewables [24].

6.5.2.2 Ancillary Services

Renewable energy projects can generate additional revenue by providing **grid-balancing services**, such as:

- **Frequency Regulation**—Maintaining grid stability by adjusting output in real time.
- **Voltage Support**—Enhancing grid reliability.
- **Spinning Reserves**—Providing backup power in case of sudden demand spikes.

Battery energy storage systems (BESS) are increasingly used to enable **solar and wind projects** to participate in ancillary service markets, further diversifying revenue streams [25].

6.5.3 *Renewable Energy Certificates (RECs)*

6.5.3.1 Understanding RECs

Renewable Energy Certificates (RECs) represent the environmental attributes[3] of electricity generated from renewable sources. These certificates can be traded separately from the physical electricity, allowing companies to meet sustainability goals [26].

Each REC typically represents **1 megawatt-hour (MWh) of renewable electricity** and can be sold to corporations, utilities, or compliance markets under Renewable Portfolio Standards (RPS).

Revenue Significance of RECs

- Voluntary and compliance markets create demand for RECs, adding a financial incentive for renewables.
- Some projects rely on REC sales as a primary revenue source, especially in markets with strong policy support.

Challenges in REC Markets

- REC prices fluctuate based on supply-demand dynamics and policy changes.
- Some markets face **double counting and verification concerns** [27].

6.5.3.2 Bundled vs. Unbundled RECs

- **Bundled RECs**—Sold together with the associated electricity under PPAs.
- **Unbundled RECs**—Sold separately from the electricity, allowing renewable projects to monetize environmental attributes independently.

Beyond understanding revenue streams like RECs, it is equally important to consider how the financial dynamics of renewable energy projects compare to those of traditional fossil fuel projects. The following section highlights key distinctions and similarities that impact project financing and risk management.

6.5.4 *Financial Model Comparison: Renewable vs. Fossil Fuel Projects*

Renewable energy projects and fossil fuel projects exhibit distinct financial characteristics due to differences in operational costs, regulatory frameworks, and market risks. Fossil fuel projects often rely on fuel price volatility and commodity

[3] Environmental attributes refer to the emissions benefits and sustainability value associated with producing electricity from renewable sources.

markets, which can lead to unpredictable cash flows but also offer opportunities for hedging. Their revenue models frequently incorporate long-term contracts or merchant exposure tied to fuel prices.

In contrast, renewable projects typically have higher upfront capital costs but significantly lower operating expenses, as fuel (wind, sunlight) is free. Consequently, revenue stability often depends on Power Purchase Agreements (PPAs) or subsidies, which reduce exposure to market price fluctuations. Additionally, renewables benefit from diversified revenue streams such as Renewable Energy Certificates (RECs) and participation in ancillary service markets.

These differences influence financing structures: fossil projects may require more flexible financing to manage price risks, while renewables attract investors favoring stable, contract-backed cash flows. Understanding these distinctions is crucial for developing balanced revenue strategies and assessing project viability in an evolving energy landscape.

By appreciating these differences in revenue models and financial risks, developers can better tailor their strategies to enhance project viability and investor confidence.

6.5.5 Conclusion

The choice of revenue model plays a **critical role** in shaping the financial sustainability of renewable energy projects. While **merchant markets** offer flexibility and higher revenue potential, they come with greater risks. **PPAs** provide stability but may limit revenue upside. Additionally, **capacity markets, ancillary services, and REC sales** offer diversification opportunities for developers. A **balanced revenue strategy**, considering a mix of these models, can enhance project viability and investor confidence.

References

1. Lazard, Lazard's levelized cost of energy analysis – version 15.0 (2022). Accessed 6 May 2025
2. A. Saltelli, S. Tarantola, F. Campolongo, *Global Sensitivity Analysis: The Primer* (Wiley, Hoboen, 2008)
3. R.A. Brealey, S.C. Myers, F. Allen, *Principles of Corporate Finance* (McGraw-Hill, New York, 2020)
4. K. Cory, B. Speer, M. Mendelsohn, *Power Purchase Agreement Checklist for State and Local Governments* (National Renewable Energy Laboratory (NREL), Masdar City, 2009)
5. F. Egli, B. Steffen, T.S. Schmidt, Finance and investment in renewable energy infrastructure. Renew. Energy **142**, 407–416 (2019)
6. R. Miller, D. Lessard, *The Strategic Management of Large Engineering Projects: Shaping Institutions, Risks, and Governance* (MIT Press, Cambridge, 2001)
7. IRENA, *Innovation Outlook: Offshore Wind* (International Renewable Energy Agency, Masdar City, 2022)

8. R. Wiser, et al., Monitoring and Adaptive Management in Renewable Energy Projects. Energy Policy **135**, 111045 (2021)
9. U.S. Department of Energy, Transmission line planning and approval challenges (2023). Accessed 27 April 2023
10. M. Bolinger, R. Wiser, B. Hoen, D. Millstein, J. Rand, G. Barbose, N. Darghouth, W. Gorman, S. Jeong, A. Mills, B. Paulos, Wind Energy Technology Data Update: 2020 Edition (webinar slide deck) (Lawrence Berkeley National Laboratory, Berkeley, 2020). https://emp.lbl.gov/sites/default/files/emp-files/bolinger_webinar_december_8_2020_16x9.pdf
11. J.C. Hull, *Options, Futures, and Other Derivatives* (Pearson Education, London, 2018)
12. C. Kreycik, T.D. Couture, K.S. Cory, Innovative feed-in tariff designs that limit policy costs. Technical Report NREL/TP-6A20-50225. National Renewable Energy Laboratory (NREL) (2011)
13. T.J. Chermack, Scenario Planning in Organizations: How to Create, Use, and Assess Scenarios (Berrett-Koehler, Oakland, 2011)
14. R. Wiser, S. Pickle, Financing investments in renewable energy: the role of policy design and restructuring. Energy Policy **26**(9), 687–707 (1998)
15. M. Bolinger, R. Wiser, W. Golove, Quantifying the value that wind power provides as a hedge against volatile natural gas prices. Technical Report LBNL-50484, Lawrence Berkeley National Laboratory (2006)
16. L. Hirth, The market value of variable renewables: the effect of solar and wind power variability on their relative price. Energy Economics **38**, 218–236 (2013)
17. World Bank, *Financing Renewable Energy: Options for Developing Countries* (World Bank Group, Washington, 2020)
18. D.J. Teece, Business models and dynamic capabilities. Long Range Plann. **51**(1), 40–49 (2018)
19. NC Clean Energy Technology Center, Database of state incentives for renewables & efficiency (dsire) (2023). https://www.dsireusa.org
20. P.L. Joskow, Challenges for wholesale electricity markets with intermittent renewable generation at scale: the us experience. Oxford Rev. Econ. Policy **35**(2), 291–311 (2019)
21. M. Brown, C. Rewey, Wholesale electricity markets and the integration of renewable energy. Technical Report, National Association of Regulatory Utility Commissioners (NARUC) (2021)
22. R.G. Newell, B.C. Prest, A. Vissing, The uncertain future of U.S. merchant renewable energy projects. Technical Report, Resources for the Future (2020)
23. Federal Energy Regulatory Commission. Energy storage and the grid (2021). https://www.ferc.gov. Retrieved from FERC
24. M. Milligan, B. Kirby, Capacity markets and variable generation. Technical report, National Renewable Energy Laboratory (NREL) (2018)
25. P. Denholm, T. Mai, R. Kenyon, Incorporating energy storage into the electric grid: storage technologies and value streams. Technical Report, National Renewable Energy Laboratory (NREL) (2019)
26. L. Bird, J. Sumner, D. Arent, Green power marketing in the united states: a status report. Technical report, National Renewable Energy Laboratory (NREL) (2021)
27. J. Heeter, M. Taylor, J. Coughlin, Status and trends in the U.S. voluntary green power market (2019 data). Technical report, National Renewable Energy Laboratory (NREL) (2020)

Chapter 7
Market Analysis and Business Strategies

7.1 Renewable Energy Market Overview

The renewable energy sector has experienced rapid growth over the past decade, fueled by increasing concerns over climate change, rising fossil fuel costs, and the global push for sustainable development. As nations strive to meet their climate goals and ensure energy security, renewable sources such as solar, wind, and battery storage have emerged as critical components of the evolving energy landscape.

7.1.1 Global Trends in Renewable Energy Adoption

Among these trends, global adoption of renewable energy technologies has accelerated significantly. The falling cost of solar and wind technologies, improved energy storage solutions, and favorable policy frameworks have catalyzed large-scale deployment across continents.

Solar Energy Solar energy remains one of the fastest-growing energy sources worldwide. The International Renewable Energy Agency (IRENA) reports that solar photovoltaic (PV) capacity grew by 18% in 2020, and it is expected to contribute nearly 10% of global electricity generation by 2030 [1]. Falling costs of solar panels due to technological advancements and economies of scale have made solar power more affordable and accessible. As of 2021, China was the largest producer of solar energy, followed by the United States and India [1].

Wind Energy Wind energy, both onshore and offshore, has also seen remarkable growth. According to the Global Wind Energy Council [2], global wind capacity increased by 53 GW in 2020 alone, with offshore wind installations growing at an unprecedented rate. Europe remains the leader in offshore wind, but other regions, including the U.S. and Asia, are increasingly investing in this technology [2].

D. Gupta et al., *The Roadmap to Renewable Energy*,
https://doi.org/10.1007/978-3-032-09609-8_7

Battery Storage The need for energy storage solutions has become more critical as the share of intermittent renewable sources like wind and solar increases. Battery storage systems, particularly lithium-ion batteries, have seen significant advances, enabling better grid management and the integration of renewables into the power grid. According to BloombergNEF [3], the global energy storage market is expected to grow by 28% annually through 2030, driven largely by the rise of electric vehicles (EVs) and renewable energy projects.

The intersection of these three sectors, solar, wind, and storage, forms the foundation for the global transition to renewable energy. Together, they represent the future of cleaner, more sustainable power generation.

7.1.2 Regional Market Analysis

The renewable energy market varies significantly across regions, driven by local resources, regulatory policies, and economic conditions. Key regions such as the U.S., Europe, and Asia have established themselves as leaders in renewable energy development, while emerging markets are catching up quickly.

United States The United States has made significant strides in renewable energy, particularly in solar and wind power. The U.S. is the second-largest solar energy market in the world, behind China, and has been a leader in wind energy, particularly in states like Texas and Iowa [4]. The passage of the Inflation Reduction Act (IRA) in 2022, which provides tax incentives for clean energy projects, has spurred even greater investments in solar, wind, and energy storage technologies [5] (American Clean Power Association, 2022). However, the long-term durability of these policy incentives remains uncertain, as future political shifts could threaten funding continuity and regulatory support.

Europe Europe has long been a frontrunner in renewable energy adoption. Wind energy, particularly offshore wind, has seen the most rapid growth. Countries like Germany, Denmark, and the United Kingdom have heavily invested in wind turbine technology, making Europe the global leader in offshore wind capacity [6]. The European Union's Green Deal, which aims for carbon neutrality by 2050, continues to drive ambitious renewable energy goals across the continent, including increasing renewable energy's share of total energy consumption to 32% by 2030 [6].

Asia Asia presents a dynamic landscape for renewable energy development. China leads the world in both solar and wind energy production, and its government is a strong proponent of clean energy, setting ambitious targets for renewable energy growth. The country has become the largest market for solar energy, producing nearly 30% of the world's solar capacity. International Renewable Energy Agency [1] is now working toward achieving 500 GW of non-fossil fuel capacity by 2030, including 280 GW from solar alone [7].The country has also launched key initiatives such as the National Green Hydrogen Mission and the Production

Linked Incentive (PLI) scheme to bolster domestic manufacturing of solar modules and energy storage technologies. Although challenges remain, including land acquisition, transmission bottlenecks, and policy implementation, India's proactive climate diplomacy, growing private-sector participation, and commitment to energy transition position it as a future global leader in renewable energy innovation. Meanwhile, Japan and South Korea are accelerating investments in offshore wind, solar, and energy storage [8], though both countries must overcome limited land availability and public opposition to certain project types. Despite these regional complexities, Asia remains a global epicenter of renewable energy innovation and capacity expansion.

Emerging Markets Africa, Latin America, and the Middle East are emerging as new frontiers for renewable energy development. In Africa, solar power is becoming increasingly popular due to its cost-effectiveness and the region's vast sun exposure. Countries like Kenya have already established themselves as leaders in geothermal energy, with geothermal power contributing significantly to Kenya's electricity grid [9]. In Latin America, Brazil and Chile are key players in the solar and wind markets, with Brazil becoming the largest wind energy producer in the region [1]. The Middle East, traditionally known for its oil wealth, is making strides in solar energy, with countries like the United Arab Emirates (UAE) investing heavily in large-scale solar projects such as the Mohammed bin Rashid Al Maktoum Solar Park [10].

7.1.3 Market Share of Leading Renewable Companies

The renewable energy industry is characterized by fierce competition, with several key players dominating the global market. These companies drive innovation, investment, and the development of large-scale renewable energy projects worldwide.

First Solar First Solar is a leading solar energy company known for its innovative thin-film photovoltaic technology. It is one of the largest producers of solar panels globally, with a market presence in the U.S., India, and other emerging markets. According to IHS Markit [11], First Solar ranks among the top solar module manufacturers globally, with a focus on utility-scale solar projects.

Vestas Vestas is a global leader in wind turbine manufacturing and installation, holding the largest market share in the global wind energy sector. The company has a strong presence in Europe, North America, and Asia and is known for its expertise in both onshore and offshore wind technologies. In 2020, Vestas had an installed capacity of more than 130 GW of wind turbines worldwide [12].

Tesla Tesla's role in the renewable energy market extends beyond electric vehicles to include energy storage and solar solutions. Tesla's Powerwall and Powerpack batteries are widely used for grid-scale energy storage and residential applications, complementing solar energy systems [13]. Additionally, Tesla's solar roof technology has gained attention for its ability to integrate solar power generation into building infrastructure.

Enel Green Power Enel Green Power, a subsidiary of the Italian multinational energy company Enel, is a major player in the global renewable energy market. With a portfolio of wind, solar, and geothermal energy projects across Europe, the Americas, and Africa, Enel Green Power is one of the largest renewable energy companies in terms of installed capacity [14].

These companies, along with others such as NextEra Energy, Siemens, Gamesa and Ørsted are driving the renewable energy transition and shaping the competitive landscape of the industry.

7.2 Market Analysis and Business Strategies

7.2.1 Business Models for Renewable Projects

As the renewable energy industry continues to grow, various business models have emerged to accommodate the unique challenges and opportunities associated with renewable energy projects. These models provide flexibility in financing, risk management, and market participation, allowing different players in the renewable sector to achieve long-term sustainability. The three key business models, Power Purchase Agreements (PPAs), Merchant Market Models, and Community Solar and Distributed Generation, are examined in this section.

7.2.2 Power Purchase Agreements (PPAs)

Power Purchase Agreements (PPAs) are long-term contracts between energy producers and purchasers (typically utilities, corporations, or government entities). These agreements outline the terms for the sale of electricity produced by renewable energy projects, such as wind, solar, or hydroelectric plants. PPAs have become a cornerstone of the renewable energy industry because they provide long-term revenue certainty, which is crucial for financing large-scale projects.

Structure of PPAs A PPA typically includes a fixed price for electricity delivered over a 10 to 25-year period, offering both the developer and the buyer financial predictability. The price may be set at a fixed rate, or it could be linked to inflation or energy market indexes [15]. For renewable projects, PPAs often include provisions for the performance of the power plant, guarantees of delivery, and terms for dispute resolution [16]. The agreement benefits the buyer by locking in energy prices at a known rate for the contract's duration, protecting against energy price volatility.

Importance of PPAs The PPA structure is essential for developers as it provides a stable revenue stream, which is necessary to secure financing from banks and investors. This stability is particularly crucial for renewable energy projects that have high capital expenditures and long payback periods. By securing a PPA, a developer can reduce the financial risk associated with energy price fluctuations and the uncertainty of future market conditions [17].

PPAs are instrumental in scaling the renewable energy sector, particularly in countries with ambitious renewable energy targets. In 2020, corporate PPAs in the U.S. reached a record 8.6 GW, driven by large companies like Google, Amazon, and Microsoft, which aim to meet their sustainability goals through renewable energy procurement [18].

7.2.3 *Merchant Market Models*

The Merchant Market Model is an alternative to PPAs, where renewable energy producers sell electricity directly to the open market, rather than entering into a long-term agreement with a purchaser. In this model, the renewable project operator takes on more market risk by exposing the revenue to fluctuations in market prices for electricity.

Risks and Benefits

1. **Risks**: The primary risk of the merchant market model is price volatility. Electricity prices can fluctuate due to factors such as changes in demand, fuel prices, and weather conditions. For instance, in times of low electricity demand or oversupply in the market, the price of power may drop significantly, leading to reduced revenues for the producer [19]. Additionally, merchant market projects are more vulnerable to policy changes, such as the removal of subsidies or regulatory support, which can affect their profitability [20].
2. **Benefits**: On the other hand, the merchant market model offers greater flexibility and potential for profit during periods of high demand or high market prices. With no long-term contract obligations, the operator can capitalize on favorable market conditions, particularly if renewable energy prices are competitive with fossil fuels. Furthermore, merchant market projects tend to have lower transaction costs compared to PPAs, as there is no need to negotiate and manage long-term contracts [21]. For developers who are able to mitigate market risks, merchant projects can offer potentially higher returns.

Some countries and regions are more conducive to merchant market models than others, depending on the level of market integration, competition, and energy demand. For instance, in markets like Spain and the U.S., where renewable energy prices are increasingly competitive, merchant models can work well for well-positioned developers [1].

7.2.4 Community Solar and Distributed Generation

Community solar and distributed generation models are changing the way renewable energy is produced and consumed, making renewable energy accessible to a wider range of consumers. These models are particularly useful in areas where large-scale renewable projects are not feasible or desirable, and they provide a means for individuals, businesses, and communities to benefit from renewable energy without the need for onsite installation.

Community Solar Community solar projects allow individuals, businesses, and municipalities to subscribe to a portion of a solar array and receive credit for the power generated, typically through a net metering arrangement. These projects are especially valuable for those who may not have the means or opportunity to install their own solar panels, such as renters or those with unsuitable rooftops. According to the National Renewable Energy Laboratory [16], community solar programs have the potential to reach 20 million households in the U.S., contributing to a more inclusive transition to renewable energy.

Revolutionizing Local Energy Markets Community solar is helping to democratize energy access and foster local ownership of renewable energy. Local governments and utilities can use these projects as a tool to meet renewable energy targets and reduce greenhouse gas emissions. In addition, community solar projects can provide direct financial benefits to local economies, creating jobs in installation, operations, and maintenance [22].

Moreover, these models promote distributed generation, where energy is generated close to where it is consumed. This decentralization reduces transmission losses, lowers energy costs, and enhances energy security by diversifying the sources of electricity. Community solar and distributed generation have the potential to revolutionize energy markets by shifting the balance of power from centralized utilities to local consumers and producers.

Distributed Generation Distributed generation refers to smaller-scale energy systems typically located close to the point of use, such as rooftop solar panels, small wind turbines, or combined heat and power (CHP) systems. These technologies allow consumers to generate their own electricity, reducing reliance on centralized grids and potentially lowering energy bills. The economics of distributed generation have become increasingly favorable: for instance, the global average cost of residential solar photovoltaics fell by more than 85% between 2010 and 2022, while the levelized cost of electricity (LCOE) from rooftop solar is now competitive with or even lower than retail electricity prices in many markets [23]. In contrast, traditional fossil fuel-based electricity generation often faces volatile input costs and growing regulatory pressures, making distributed generation an attractive alternative for energy security and affordability [23].

Moreover, distributed generation enhances system resilience by creating decentralized networks that are less vulnerable to widespread outages caused by natural disasters or grid failures. As renewable technologies continue to improve and supportive policy frameworks expand, distributed generation is expected to play an increasingly central role in accelerating the global energy transition [23].

In addition, distributed generation enhances the resilience of energy systems by creating decentralized energy grids that are less vulnerable to natural disasters or technical failures. As technology advances and the costs of renewable energy systems continue to decline, distributed generation will play an increasingly important role in the global energy transition [10].

7.3 Market Analysis and Business Strategies

7.3.1 Competitive Positioning and Market Dynamics

In the renewable energy sector, achieving a competitive edge requires companies to navigate the complex dynamics of cost competitiveness, market differentiation, and regulatory influences. These factors shape the ability of companies to thrive in an increasingly crowded and competitive market. This section explores the competitive positioning of renewable energy projects, with a focus on cost reductions in solar and wind, strategies for market differentiation, and the role of policy and regulatory frameworks.

7.3.1.1 Cost Competitiveness in Renewable Energy

Renewable energy sources, particularly solar and wind, have become significantly more cost-competitive in recent years. The reduction in the cost of these technologies is due to various factors, including technological innovations, economies of scale, and improved supply chain management. These factors have allowed renewables to become more affordable, and in many regions, they now represent the lowest-cost source of new power generation (Fig. 7.1).

Factors Driving Cost Reductions in Solar and Wind

1. **Economies of Scale**: The cost reductions in solar and wind energy have largely been driven by economies of scale. As both industries have expanded, manufacturers have been able to lower their production costs due to increased volume. The deployment of large-scale solar and wind farms, in turn, reduces costs for individual installations [10]. As the renewable energy industry has matured, both supply chain efficiencies and more competitive pricing in materials, such as steel for wind turbines and silicon for solar panels, have also played a significant role in reducing costs [25].
2. **Technological Advancements**: Technological innovations in solar panels and wind turbines have improved efficiency and performance while lowering costs. For instance, the development of bifacial solar panels, which capture sunlight

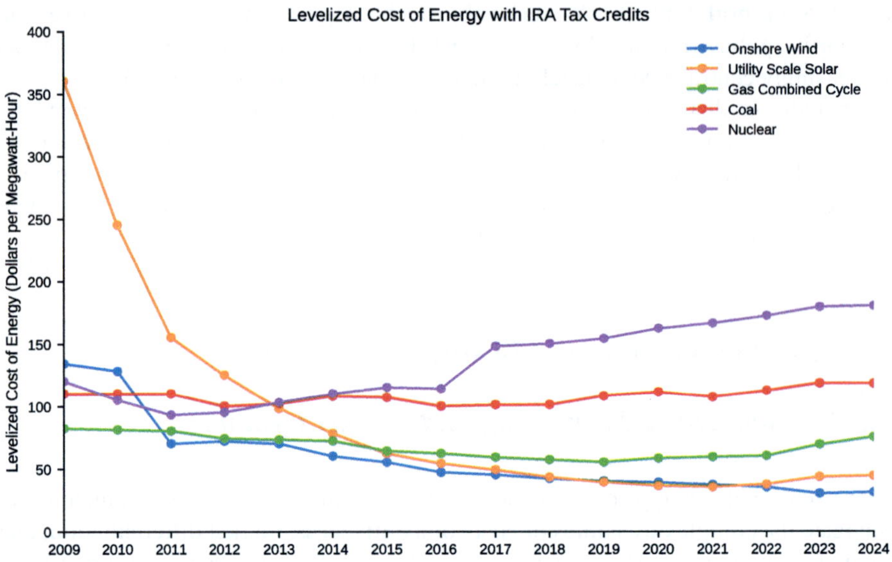

Fig. 7.1 U.S. electricity levelized costs over time, including IRA clean energy tax credits [24]

on both sides, and taller wind turbine towers have contributed to greater energy output and lower operational costs [10]. The continuous improvements in turbine size and blade technology, for example, have led to higher capacity factors for wind farms, making wind energy more efficient and cost-effective [3].

3. **Declining Financing Costs**: As the renewable energy market matures, financing for renewable projects has become more affordable. The decreasing cost of capital, driven by investor confidence in the sector and the increasing role of green financing, has made it easier for renewable energy projects to secure the capital necessary for development [25]. Additionally, favorable financing models such as power purchase agreements (PPAs) allow developers to secure long-term contracts at favorable rates, further reducing financial risk.

According to the International Renewable Energy Agency [10], between 2010 and 2020, the global levelized cost of electricity (LCOE) for utility-scale solar PV declined by 89%, while the LCOE for onshore wind dropped by 70%. This dramatic cost reduction has made renewable energy sources more attractive than conventional fossil fuel-based power generation in many regions.

7.3.1.2 Strategies for Market Differentiation

As competition intensifies in the renewable energy sector, companies are increasingly looking for ways to differentiate themselves. Effective differentiation strategies can help companies gain a competitive edge, attract customers, and secure investment. Companies that excel in innovation, operational efficiency, and customer engagement are well-positioned to lead the market.

Innovation in Technology Companies are leveraging advanced technologies to offer more efficient solutions, reduce operational costs, and enhance energy production. For instance, wind turbine manufacturers are continuously improving turbine designs to increase output, while solar developers are exploring hybrid models that combine wind and solar power for enhanced generation. Companies that adopt cutting-edge technologies and stay ahead of industry trends are more likely to differentiate themselves in a competitive marketplace [1].

Operational Efficiency Companies that streamline their operations, reduce costs, and improve the efficiency of energy generation are better positioned to outperform competitors. For example, renewable energy companies that integrate predictive maintenance technology to monitor the health of wind turbines and solar panels can lower downtime and improve system reliability, leading to higher returns on investment [3].

Customer-Centric Approaches In addition to technological and operational advancements, successful renewable energy companies differentiate themselves by offering tailored solutions that meet customer needs. For example, offering flexible renewable energy procurement options, such as community solar programs or corporate renewable energy procurement options like power purchase agreements (PPAs), enables companies to attract both residential and corporate customers [10].

7.3.2 Policy and Regulatory Impacts on Market Dynamics

Government policies and regulations play a critical role in shaping the competitive dynamics of the renewable energy market. Policies that support renewable energy development can level the playing field for clean energy projects and encourage investment, while regulatory challenges can slow down or impede market growth. It is essential for companies to stay informed about current and future regulatory changes to maintain their competitive positioning.

Incentives and Subsidies Policies such as tax credits, feed-in tariffs, and renewable portfolio standards have been instrumental in driving the growth of the renewable energy market. In the U.S., the Investment Tax Credit (ITC) for solar and the Production Tax Credit (PTC) for wind energy have played key roles in lowering upfront costs for developers and accelerating market penetration [25]. These incentives reduce the financial barriers to entry for renewable energy projects and make it more feasible for developers to bring projects online.

Carbon Pricing and Emission Regulations Carbon pricing mechanisms, such as carbon taxes or cap-and-trade systems, can also significantly impact market competition. These policies encourage the transition from fossil fuel-based generation to renewables by making carbon-intensive power generation more expensive. The introduction of carbon pricing mechanisms can create more favorable market conditions for renewable energy, further driving down the cost-competitiveness gap between clean and traditional energy sources [1].

Policy Uncertainty While supportive policies have driven renewable energy growth, policy uncertainty can have the opposite effect. Changes in government leadership, policy priorities, or regulatory frameworks can disrupt market dynamics and investment in renewable projects. For instance, the rollback of renewable energy subsidies or changes in energy efficiency standards can undermine long-term investment strategies and slow down industry growth [3].

The evolving regulatory landscape presents both opportunities and challenges for companies in the renewable energy sector. By staying ahead of regulatory changes and adapting to new policy environments, companies can effectively navigate the shifting market dynamics.

7.4 Growth Strategies and M&A in Renewable Energy

The renewable energy industry has experienced significant growth in recent years, driven by global shifts toward sustainability, technological advancements, and policy support. Companies in the sector have increasingly adopted various growth strategies to expand their reach, enhance competitiveness, and drive profitability. Among these strategies, mergers and acquisitions (M&A), scaling renewable portfolios, and strategic partnerships play a critical role in the industry's development. This chapter explores the key aspects of M&A trends, scaling renewable portfolios globally, and the value of strategic partnerships and joint ventures in accelerating growth.

7.4.1 Mergers and Acquisitions

Mergers and acquisitions have become pivotal tools for accelerating growth in the renewable energy sector. In recent years, there has been a noticeable uptick in M&A activity as companies look to consolidate their positions, expand their renewable energy assets, and integrate new technologies. This trend is particularly evident in the solar, wind, and battery storage markets, where large energy corporations and utilities are acquiring smaller, innovative companies to diversify their portfolios and enhance operational efficiency.

The impact of M&A on the renewable energy industry is multi-faceted. On one hand, it enables companies to rapidly increase their scale, gain access to new markets, and optimize their project development capabilities. For example, the acquisition of SunPower by TotalEnergies in 2011 helped TotalEnergies to integrate solar energy solutions into its global portfolio, marking a significant shift toward renewable energy [26]. On the other hand, M&A can foster the adoption of cutting-edge technologies, as seen with NextEra Energy's acquisition of wind turbine manufacturers, allowing for more cost-effective and efficient energy production [27].

Recent M&A trends indicate a shift toward the strategic acquisition of companies with strong development pipelines, which reduces the time to market for renewable energy projects. This trend is expected to continue as firms seek to enhance their competitive edge and adapt to regulatory changes that support renewable energy transition.

7.4.2 Scaling Renewable Portfolios

Scaling renewable energy portfolios is a vital strategy for companies looking to expand their global footprint and capitalize on the growing demand for clean energy. Companies pursuing this strategy typically focus on increasing the number and diversity of renewable projects within their portfolios, ranging from solar and wind to emerging technologies such as geothermal and hydrogen.

To scale effectively, energy companies must employ several strategies. First, they often target geographic expansion into regions with strong growth potential, driven by supportive policies and favorable resource availability. For instance, countries in Asia and Africa, with their abundant solar and wind resources, are becoming hotspots for international renewable energy investments [28]. Companies with established success in one region are increasingly looking to replicate that success in emerging markets, where renewable energy adoption is still in its early stages.

Second, scaling renewable portfolios involves integrating various technologies that complement each other, providing a more resilient energy mix. For example, integrating solar and battery storage systems enables better load balancing and energy reliability, creating a competitive advantage in markets where grid stability is a concern [10]. Scaling also includes enhancing supply chain capabilities, improving financing structures, and establishing efficient project development teams that can deploy multiple projects simultaneously.

7.4.3 Strategic Partnerships and Joint Ventures

Strategic partnerships and joint ventures have become integral in fostering innovation, expanding renewable energy portfolios, and accelerating the deployment of new technologies. By collaborating with other industry players, companies can pool resources, share risks, and access new markets more efficiently.

One of the key benefits of strategic partnerships is the ability to share expertise and resources. For example, in offshore wind projects, partnerships between renewable energy companies and oil and gas majors have been instrumental in addressing the technical and financial challenges of these large-scale projects [3]. The collaboration between Ørsted and BP for offshore wind development in the United States is a prime example, where both companies leverage each other's expertise in offshore infrastructure and financing to deploy large-scale projects [21].

Moreover, joint ventures allow companies to mitigate risks associated with new markets and technologies. In the rapidly evolving renewable energy sector, where regulatory frameworks and technological advances are constantly changing, entering a joint venture allows companies to take a more cautious, measured approach while gaining access to innovative solutions. The collaboration between Tesla and Panasonic in the development of battery storage technologies is a notable example of how partnerships can foster innovation while simultaneously driving business expansion [29].

In the future, as the demand for renewable energy continues to rise, the importance of strategic partnerships and joint ventures will only increase. Companies that successfully leverage these collaborations will be better positioned to scale their operations globally, reduce operational risks, and stay at the forefront of technological advancements in the energy sector.

In short, the renewable energy sector's growth trajectory will largely depend on how companies navigate key strategies such as mergers and acquisitions, scaling portfolios, and forming strategic partnerships. M&A activity will continue to drive consolidation and the integration of new technologies, while scaling renewable energy portfolios across diverse geographies will enable companies to capitalize on global demand. Finally, strategic partnerships and joint ventures will play a critical role in accelerating growth, driving innovation, and expanding market reach. Companies that effectively leverage these growth strategies will be well-positioned to thrive in the evolving energy landscape.

7.5 Emerging Markets and Opportunities

The global shift toward renewable energy is creating substantial opportunities in emerging markets, where demand for clean energy is growing rapidly due to increasing populations, industrialization, and urbanization. Latin America, Southeast Asia, and Africa are three regions where the potential for renewable energy development is immense. In these areas, local governments, international investors, and energy companies are exploring ways to tap into the vast renewable energy potential. This chapter delves into the renewable energy trends and opportunities within these emerging markets, highlighting key regions and the associated challenges.

7.5.1 Renewable Energy in Latin America

Latin America is experiencing a renewable energy revolution, driven by a combination of favorable geographic conditions, supportive government policies, and growing demand for sustainable energy solutions. The region is rich in renewable resources, including solar, wind, hydro, and geothermal energy. As a result, several Latin American countries have become leaders in the adoption of renewable energy, particularly in the solar and wind sectors.

Brazil is one of the largest renewable energy markets in the region, with a well-established wind energy industry and a rapidly growing solar sector. The country has invested heavily in wind farms, particularly in the northeast, where strong winds create ideal conditions for power generation. In addition to wind, Brazil's solar energy market has been expanding rapidly in recent years, spurred by declining technology costs and strong government incentives (International Renewable Energy Agency [1]. The country's renewable energy sector benefits from a robust regulatory framework that includes long-term power purchase agreements (PPAs) and net metering policies that support private investment.

Chile is another standout in Latin America, particularly in solar energy. With one of the best solar resources globally, Chile has become a hub for solar energy development. The Atacama Desert, with its intense solar radiation, offers a prime location for solar power projects. Chile's commitment to a carbon-neutral future by 2050 has further accelerated the country's efforts to increase the share of renewables in its energy mix. Chile also introduced a significant renewable energy auction system, encouraging private investments in clean energy [30].

Argentina and Mexico are also emerging players in Latin America's renewable energy transition, focusing on wind, solar, and geothermal projects. Mexico, with its strong policy push for energy reform, has attracted significant foreign investments in renewable energy. Argentina, despite economic challenges, has been advancing its renewable energy projects, primarily in the wind energy sector, aided by international financial support and partnerships.

7.5.2 Southeast Asia: A Growing Renewable Hub

Southeast Asia has quickly become one of the world's fastest-growing renewable energy markets. The region's energy demand is expected to increase by over 60% by 2040 due to rapid economic growth and urbanization. As the demand for energy grows, Southeast Asian countries are looking to expand their renewable energy capacity to meet sustainability goals and reduce dependency on fossil fuels.

Countries such as Vietnam, Indonesia, and the Philippines are leading the charge in renewable energy development in Southeast Asia. Vietnam is one of the fastest-growing solar markets in the world, driven by government support, an attractive feed-in tariff (FiT) system, and a booming private sector. Solar power capacity in the country surged dramatically in recent years, making it a regional leader in renewable energy adoption [8].

Indonesia, with its vast geothermal potential, is also positioning itself as a key player in the renewable energy transition. The country has the largest geothermal resources in the world and has begun to harness this potential for power generation. Indonesia's government has outlined plans to increase its share of renewables in its energy mix to 23% by 2025, with a significant focus on geothermal, hydropower, and solar energy [9].

The Philippines has made considerable progress in expanding its renewable energy capacity, particularly in wind and solar. The country's geography provides excellent conditions for wind energy, with many projects being developed in coastal areas. Additionally, the government has set ambitious renewable energy targets, including the goal to increase the share of renewables in the power generation mix to 35% by 2030 [1].

Despite these successes, there are several challenges to renewable energy growth in Southeast Asia. These include regulatory barriers, insufficient grid infrastructure, and financing constraints. Furthermore, the region's reliance on coal and natural gas for electricity generation creates a complex political and economic landscape for renewable energy adoption. Nevertheless, with continued policy support, technological advancements, and international investment, Southeast Asia has the potential to become a global renewable energy leader.

7.5.3 Africa's Renewable Energy Potential

Africa is home to vast renewable energy resources that remain largely untapped. The continent has an abundance of solar, wind, hydro, and geothermal energy potential, yet access to electricity remains a significant challenge for many African countries. Despite this, renewable energy development is gaining momentum, with several African nations aiming to transition to sustainable energy systems.

Solar energy holds the most promise for Africa, given the region's vast sun exposure. Sub-Saharan Africa, in particular, is ideal for solar power generation, with several countries receiving high levels of solar radiation year-round. Countries such as Kenya, South Africa, and Morocco are already leading the way in solar energy development. Morocco's Noor Ouarzazate Solar Complex, one of the largest solar power plants in the world, exemplifies Africa's potential to harness solar energy at a large scale [10]. In South Africa, the government has rolled out the Renewable Energy Independent Power Producer Procurement Programme (REIPPPP), which has attracted significant investments in solar and wind energy projects.

Wind energy also presents a promising opportunity, particularly in countries along the eastern and southern coasts. South Africa and Egypt have developed substantial wind farms, and the potential for further growth in wind energy is significant. In East Africa, Kenya is emerging as a leader in wind energy, with the Lake Turkana Wind Power Project being one of the largest wind farms in Africa [31].

Geothermal energy is another underutilized resource in Africa. The East African Rift Valley, stretching from the Red Sea to Mozambique, contains some of the world's richest geothermal resources. Kenya has already made significant strides in tapping into this potential, becoming a regional leader in geothermal power generation. Ethiopia is also developing geothermal energy projects, and other countries such as Tanzania and Uganda are exploring the feasibility of harnessing this resource.

Despite the enormous potential, Africa faces significant challenges in its renewable energy transition, including limited access to capital, political instability, and inadequate infrastructure. However, international organizations, development banks, and private investors are increasingly looking to support renewable energy projects in Africa. The African Development Bank's New Deal on Energy for Africa aims to bring electricity to millions of people across the continent and is expected to play a key role in accelerating the adoption of renewable energy [31].

7.5.4 Conclusion

Emerging markets in Latin America, Southeast Asia, and Africa offer substantial growth opportunities for the global renewable energy sector. Each region has unique advantages, including abundant renewable resources, favorable government policies, and increasing demand for sustainable energy. However, challenges such as regulatory hurdles, financing constraints, and inadequate infrastructure must be overcome to fully realize the potential of renewable energy. As the global energy transition accelerates, these regions are poised to play a central role in shaping the future of renewable energy.

References

1. International Renewable Energy Agency (IRENA), Renewable capacity statistics (2021)
2. Global Wind Energy Council (GWEC), Global wind report 2020 (2020). https://gwec.net/
3. BloombergNEF, Battery pack prices fall as market ramps up with market average at $132/kwh in 2021 (2021)
4. U.S. Department of Energy, Renewable energy (2021). Accessed 09 May 2025
5. American Wind Energy Association, Wind energy in texas (2022)
6. European Commission, Renewable energy directive (2020)
7. Ministry of New and Renewable Energy, National solar mission (2023). https://mnre.gov.in
8. Asian Development Bank, Asia's renewable energy transition: opportunities and challenges (2021). Accessed 09 May 2025
9. World Bank, Scaling up to phase down: financing energy transitions in developing countries (2021). Accessed 09 May 2025
10. IRENA, *Renewable Power Generation Costs in 2019* (International Renewable Energy Agency, Masdar City, 2020)

11. IHS Markit, Global pv market outlook: first solar among top utility-scale module suppliers (2021). Accessed 09 May 2025
12. Vestas, Annual report 2020 (2021). Accessed 09 May 2025
13. Inc. Tesla, 2020 impact report (2021). Accessed 09 May 2025
14. Enel Green Power, Annual report 2020 (2021). Accessed 09 May 2025
15. T. Hirsch, Understanding power purchase agreements: key structures and terms. *Renewable Energy World* (2021). Accessed 09 May 2025
16. National Renewable Energy Laboratory (NREL), A guide to power purchase agreements for renewable energy projects (2021). Accessed 09 May 2025
17. S&P Global, The role of power purchase agreements in renewable energy financing (2020). Accessed 09 May 2025
18. American Clean Power Association, Corporate renewable energy procurement reaches record high in 2020 (2020). Accessed 09 May 2025
19. U.S. Energy Information Administration (EIA), Electricity prices and market trends in 2020 (2020). Accessed 09 May 2025
20. P. Caldwell, J. Smith, S. Lee, Regulatory risks in the merchant market model: a case study of renewable energy. Energy Econ. **94**, 122–130 (2021)
21. Deloitte, The benefits of the merchant market model for renewable energy developers (2021). Accessed 09 May 2025
22. C. Hernandez, E. Smith, R. Johnson, Revolutionizing local energy markets: the role of community solar in democratizing energy access. Renew. Energy J. **89**, 25–35 (2020)
23. International Renewable Energy Agency, Renewable power generation costs in 2022 (2023)
24. I. Orr, M. Rolling, Your guide to lazard's lcoe analysis part 4: Lazard says wind and solar costs are increasing (2023). Accessed 4 Aug 2025
25. Lazard, Levelized cost of energy analysis – version 14.0 (2020). Accessed 09 May 2025
26. S. Carter, The role of mergers and acquisitions in accelerating renewable energy adoption. Renew. Energy J. **45**(2), 134–142 (2019)
27. McKinsey & Company, Global energy perspective 2022: accelerating the energy transition (2022). Accessed 09 May 2025
28. KPMG International, Renewable energy: growth and investment opportunities in emerging markets (2021). Accessed 10 May 2025
29. Tesla Inc., Tesla and panasonic expand partnership for battery innovation (2020). Accessed 10 May 2025
30. SolarPower Europe, Global market outlook for solar power 2020–2024 (2020). https://www.solarpowereurope.org/insights/global-market-outlook-for-solar-power-2020-2024. Accessed 10 May 2025
31. African Development Bank, Lake turkana wind power project (2021). https://www.afdb.org/en/projects-and-operations/p-ke-fd0-002. Accessed 10 May 2025

Chapter 8
Data Science and Machine Learning in Renewable Energy

8.1 Introduction

The renewable energy sector is undergoing rapid changes as the world transitions toward sustainable energy sources. Technologies like solar photovoltaic panels, wind turbines, hydroelectric systems, and battery storage are becoming more sophisticated. However, alongside technological growth, another powerful force is shaping the future of renewable energy: data.

Every solar farm, wind farm, and smart grid today generates massive amounts of data. This includes measurements like energy production, weather conditions, equipment status, and customer usage patterns. Properly analyzing and utilizing this data can dramatically improve how renewable energy systems are designed, operated, and maintained.

Renewable energy practitioners can greatly benefit from using **Data Science** (DS) and **Machine Learning** (ML) for these problems. Data Science is the process of extracting useful insights from data. Machine Learning, a subset of Data Science, involves teaching computers to learn patterns from historical data and make predictions about the future.

For professionals in the renewable energy field, learning how to apply basic data science and machine learning techniques can unlock new opportunities. These include:

- Predicting how much energy a solar farm will generate next month.
- Forecasting future electricity demand in a city.
- Identifying when a wind turbine might need maintenance before it fails.
- Estimating future revenue from a renewable energy project.

This chapter is aimed to function as a step-by-step guide to understand how Machine Learning can be used to solve business problems in this domain.

D. Gupta et al., *The Roadmap to Renewable Energy*,
https://doi.org/10.1007/978-3-032-09609-8_8

8.2 Machine Learning Lifecycle: A Step-by-Step Guide

Machine learning is not a single tool but a step-by-step process. Like building a solar farm or a wind turbine, machine learning projects have a lifecycle: from identifying a need, gathering materials, building, testing, and finally operating and maintaining the system. In this section, we walk through the full lifecycle of a machine learning project, focusing on renewable energy applications.

8.2.1 Problem Formulation

Every successful project begins by clearly defining the problem. In renewable energy, typical business needs might be:

- How much energy will our solar farm produce in the next 12 months?
- How much electricity demand will we see in a particular city during the upcoming winter?
- Will a particular wind farm meet its financial revenue targets next year?

Machine learning requires problems to be structured in a specific way, often by asking: **What do we want to predict?**

Two common types of machine learning problems are:

- **Regression Problems**: Predicting a continuous quantity, such as:
  - How many megawatt-hours (MWh) of energy a wind farm will produce next month.
  - The expected revenue (in dollars) from a solar project next year.
- **Classification Problems**: Predicting a category or label, such as:
  - Whether a battery storage system will need maintenance within the next six months (Yes or No).

Why Is This Step Important? Clearly formulating the problem ensures that the correct type of data is collected, the right kind of model is chosen, and the evaluation is meaningful. If this step is rushed or unclear, the entire project can fail even if the technical work later is excellent.

Example Suppose you want to predict how much electricity a residential solar installation will produce each month. This is a **regression problem** because energy production is a continuous number (e.g., 5000 kWh per month). If you instead want to predict whether the solar system will meet its monthly production guarantee (Yes/No), it becomes a **classification problem**.

8.2.2 Data Collection and Integration

Once the problem is clearly defined, the next step is to gather data that could help solve the problem.

Useful data sources include:

- **Historical Energy Data**: Records of energy generated by solar farms, wind farms, or hydro plants over time.
- **Weather and Meteorological Data**: Information on sunlight (solar irradiance), wind speeds, temperature, humidity, and rainfall.
- **Market and Financial Data**: Data on electricity prices, renewable energy certificate (REC) markets, tariffs, and financial incentives.
- **Asset Maintenance Records**: Logs detailing equipment maintenance, repair, and replacement.

Public datasets include:

- **NREL End-Use Load Profiles**: Hourly electricity usage profiles across U.S. building types.
- **National Solar Radiation Database (NSRDB)**: Solar and meteorological data for North America.
- **NOAA Weather Data**: Historical and forecasted weather conditions.

Why Is This Step Important? Machine learning models are only as good as the data they are trained on. High-quality, relevant, and clean data greatly improves model performance.

Example Predicting next month's solar farm energy output can be improved by including average solar irradiance, number of rainy days, and temperature variations.

8.2.3 Data Cleaning and Preprocessing

After gathering raw data, the next step is preparing it for machine learning models. Real-world data is often messy, incomplete, inconsistent, or noisy.

Key tasks include:

1. **Handling Missing Data**: Measurements may sometimes be missing, such as when an energy meter fails to record data.
   - **Imputation**: Fill missing values using the mean, median, or interpolated values.
   - **Removal**: Drop rows with missing values when appropriate.
2. **Removing Outliers**: Outliers are abnormal values that do not make sense (e.g., zero energy generation on a sunny day). Detecting and removing outliers ensures the model does not learn incorrect patterns.

3. **Normalization and Standardization**: Features may have different ranges:

   - Energy output: 0 to 10,000 kWh.
   - Temperature: −10 to 40 °C.

   Solutions:

   - **Normalization**: Rescale to [0,1].
   - **Standardization**: Rescale to zero mean and unit variance.

   Simple Intuition: Without adjusting the scale, features like energy output could dominate models unfairly.

   Example: If energy output (in thousands) and wind speed (in tens) are combined, the model may overemphasize energy output unless normalized.
4. **Resampling and Consistency**: For time-series data, ensure consistent data intervals. If some days have missing hours or inconsistent recording, fix it through resampling or interpolation.

8.2.4 Feature Engineering

Feature engineering is the process of creating new inputs for the model from the existing data. Good feature engineering often makes the difference between a mediocre model and an excellent one.

Common techniques include:

1. **Time-based Features**: Time often affects energy usage and production. Creating new features from timestamps can help:

   - **Hour of Day**: Energy usage is often higher during daytime working hours.
   - **Day of Week**: Weekends may show different energy patterns compared to weekdays.
   - **Month of Year**: Summer and winter months usually have distinct energy profiles.
   - **Holiday Indicators**: Special holidays (e.g., Christmas, Thanksgiving) might have unique consumption patterns.

2. **Lag Features**: Using the value of a variable at a previous time step to predict the current value.

   Simple Explanation: If you want to predict today's energy usage, it makes sense to look at yesterday's usage. This is called a **lag**.

   Example lag features:

   - Energy usage 24 hours ago.
   - Energy usage exactly 1 week ago.

3. **Rolling Aggregates**: Instead of looking at a single previous value, create aggregates:
   - **Rolling Mean**: Average energy usage over the past 7 days.
   - **Rolling Sum**: Total energy usage in the last month.
4. **Interaction Features**: Sometimes two features combined provide better information than individually.
   - Example: Solar irradiance × panel area = Potential energy production.

Remember Feature engineering requires domain knowledge. Your understanding of how weather, technology, and demand interact can lead to smarter feature creation.

8.2.5 Model Selection

Once the dataset is ready with clean and meaningful features, the next major step is choosing a model.

Commonly used models:

1. **Linear Regression**:
   - **What it is:** Fits a straight line to predict outputs.
   - **Simple Intuition:** Energy demand increasing with temperature can be approximated with a straight line.
   - **Pros**: Simple, fast, easy to interpret.
   - **Cons**: Cannot capture non-linear patterns.
2. **Decision Trees**:
   - **What it is:** Splits data based on conditions.
   - **Simple Intuition:** Like a flowchart to make decisions.
   - **Pros**: Captures complex relationships.
   - **Cons**: Can overfit if not controlled.
3. **Random Forests**:
   - **What it is:** Collection of decision trees with averaged predictions.
   - **Simple Intuition:** Uses "wisdom of crowds."
   - **Pros**: Powerful, accurate.
   - **Cons**: Slower to train than simple models.
4. **XGBoost**:
   - **What it is:** Advanced gradient-boosted decision trees.
   - **Pros**: High accuracy.
   - **Cons**: Harder to tune.

5. **Recurrent Neural Networks (RNN) and Long Short-Term Memory (LSTM)**:
   - **What it is:** Specialized for sequential/time-series data.
   - **Pros**: Excellent for energy demand forecasting.
   - **Cons**: Require more data and are harder to interpret.

Choosing the Right Model
- Start simple (Linear Regression or Decision Tree).
- If simple models perform poorly, try Random Forest or XGBoost.
- For sequence/time-series data, use LSTM.

8.2.6 Training and Hyperparameter Tuning

Training involves showing the model examples from historical data.

1. **Train-Test Split**: Divide data into:
   - Training Set (70–80%): For learning.
   - Testing Set (20–30%): For evaluation.
2. **Cross-Validation**: Splits data multiple times to ensure stable results.
   Simple Intuition: Like testing solar panels under different weather conditions.
3. **Hyperparameter Tuning**: Improves model performance.
   - **Examples**: Tree depth, number of trees.
   - **Methods**: Grid Search, Random Search, Bayesian Optimization.

8.2.7 Model Evaluation

Systematic evaluation quantifies model performance.

1. **Mean Absolute Error (MAE)**:

$$\text{MAE} = \frac{1}{n}\sum_{i=1}^{n} |y_i - \hat{y}_i|$$

   Simple average of absolute errors.
2. **Root Mean Squared Error (RMSE)**:

$$\text{RMSE} = \sqrt{\frac{1}{n}\sum_{i=1}^{n}(y_i - \hat{y}_i)^2}$$

   Penalizes larger errors more.

3. **Mean Absolute Percentage Error (MAPE)**:

$$\text{MAPE} = \frac{100\%}{n} \sum_{i=1}^{n} \left| \frac{y_i - \hat{y}_i}{y_i} \right|$$

   Expresses error as a percentage.

Business Connection An RMSE of 500 kWh might require an extra $10,000 battery, linking model error directly to costs.

8.2.8 Deployment and Monitoring

After successful training and evaluation, deploy the model.

1. **Deployment**: Make the model available through:
   - Dashboards for plant managers.
   - Automated energy storage control.
2. **Monitoring**: Maintain model accuracy over time.
   - Retrain with new data.
   - Detect concept drift due to climate changes or technology upgrades.

Just like maintaining physical renewable systems, machine learning models require continuous maintenance to stay effective.

8.3 Hands-On Case Study: Power Forecasting with Machine Learning

To complement the theoretical walkthrough of the machine learning lifecycle, this section presents a hands-on case study focused on forecasting electricity demand in Tetouan City, Morocco [1]. Using real-world weather and power consumption data, we demonstrate how each step in the ML pipeline, ranging from data cleaning and feature engineering to model training and evaluation, can be applied in practice.

Two Jupyter notebooks accompany this chapter:

- A **basic version**, designed for readers who are new to machine learning, introducing core concepts and simple models.
  Colab link: https://colab.research.google.com/drive/1ylHGIiExzFHg7G2zuIXzRph60HKb5GDA
- An **advanced version**, aimed at more experienced readers, featuring richer feature engineering and improved model performance using more sophisticated techniques.

Colab link: https://colab.research.google.com/drive/1wavdTSwnwQj_riIguoN9STHH7Qu_Jy7-

Readers are strongly encouraged to open and run these notebooks in Colab to experiment with the code, modify parameters, and gain deeper insights into how machine learning can be applied to renewable energy problems.

For convenience and offline access, static PDF versions of the notebooks are also included in this book. These serve as a reference for readers who may be using a printed copy or working in environments without internet access.

Appendix: Power Consumption Forecasting for Tetouan City

Motivation and Business Problem

This notebook aims to serve as a quickstarter guide to demonstrate how Machine Learning (ML) techniques can be applied to real-world problems in the renewable energy sector.

Context

Tetouan, a city in northern Morocco with a population of approximately 380,000, is undergoing rapid urbanization, leading to increased and variable energy demands. Power utilities and energy planners require data-driven strategies to efficiently manage resources and ensure uninterrupted electricity supply.

Accurate forecasting of power consumption at a granular level enables:

- Efficient energy distribution
- Reduced operational costs
- Improved reliability of power infrastructure
- Better planning for future energy infrastructure investments

The dataset used in this notebook is sourced from the UCI Machine Learning Repository, which provides zone-wise power consumption and meteorological data for Tetouan City.

Objective

We demonstrate how ML & AI can aid in forecasting electricity consumption in order to efficiently adapt to variable energy demands. By anticipating usage patterns, we can strike a balance between meeting growing energy needs and minimizing energy wastage. This involves:

- Analyzing consumption patterns over time
- Identifying key environmental and temporal drivers of electricity use
- Developing predictive models using historical power usage and weather data

By achieving these goals, the notebook aims to showcase how AI can enable smarter grid operations and more sustainable energy planning.

Data Loading and Initial Setup

[17]:
```
#! pip install -r requirements.txt
```

[18]:
```
import pandas as pd
import numpy as np
import matplotlib.pyplot as plt
import seaborn as sns
import warnings
warnings.filterwarnings('ignore')
import requests
import zipfile
import io
from sklearn.linear_model import LinearRegression, Ridge, Lasso
from sklearn.ensemble import RandomForestRegressor, GradientBoostingRegressor
from sklearn.metrics import mean_absolute_error, mean_squared_error, r2_score

# Set random seed for reproducibility
np.random.seed(42)

# Set better plot aesthetics
plt.style.use('seaborn-whitegrid')
sns.set_palette("viridis")
```

[19]:
```
# Download and extract CSV from ZIP
url = "https://archive.ics.uci.edu/static/public/849/
 ↪power+consumption+of+tetouan+city.zip"
response = requests.get(url)
with zipfile.ZipFile(io.BytesIO(response.content)) as z:
    with z.open(z.namelist()[0]) as f:
        df = pd.read_csv(f)

# Display the first few rows
print("Dataset initial shape:", df.shape)
df.head()
```

```
Dataset initial shape: (52416, 9)
```

[19]:

	DateTime	Temperature	Humidity	Wind Speed	general diffuse flows	\
0	1/1/2017 0:00	6.559	73.8	0.083	0.051	
1	1/1/2017 0:10	6.414	74.5	0.083	0.070	
2	1/1/2017 0:20	6.313	74.5	0.080	0.062	
3	1/1/2017 0:30	6.121	75.0	0.083	0.091	
4	1/1/2017 0:40	5.921	75.7	0.081	0.048	

```
   diffuse flows  Zone 1 Power Consumption  Zone 2  Power Consumption  \
0          0.119               34055.69620                 16128.87538
1          0.085               29814.68354                 19375.07599
2          0.100               29128.10127                 19006.68693
3          0.096               28228.86076                 18361.09422
4          0.085               27335.69620                 17872.34043

   Zone 3  Power Consumption
0                 20240.96386
1                 20131.08434
2                 19668.43373
3                 18899.27711
4                 18442.40964
```

Data Cleaning and Preprocessing

[20]:
```
# Preview structure
print("\nDataset information before preprocessing:")
df.info()
```

```
Dataset information before preprocessing:
<class 'pandas.core.frame.DataFrame'>
RangeIndex: 52416 entries, 0 to 52415
Data columns (total 9 columns):
 #   Column                     Non-Null Count  Dtype
---  ------                     --------------  -----
 0   DateTime                   52416 non-null  object
 1   Temperature                52416 non-null  float64
 2   Humidity                   52416 non-null  float64
 3   Wind Speed                 52416 non-null  float64
 4   general diffuse flows      52416 non-null  float64
 5   diffuse flows              52416 non-null  float64
 6   Zone 1 Power Consumption   52416 non-null  float64
 7   Zone 2  Power Consumption  52416 non-null  float64
 8   Zone 3  Power Consumption  52416 non-null  float64
dtypes: float64(8), object(1)
memory usage: 3.6+ MB
```

[21]:
```
# Removes extra spacing from column names for consistency
df.rename(columns={
    'Zone 2  Power Consumption': 'Zone 2 Power Consumption',
    'Zone 3  Power Consumption': 'Zone 3 Power Consumption'
}, inplace=True)

# Convert 'DateTime' column which is currently an 'object' datatype to␣
 ↪'datetime' datatype and set it as index
```

```
df['DateTime'] = pd.to_datetime(df['DateTime'], dayfirst=True)
df.set_index('DateTime', inplace=True)

# Preview structure
print("\nDataset information after preprocessing:")
df.info()

# Define key column groups for easier reference
power_cols = ['Zone 1 Power Consumption', 'Zone 2 Power Consumption', 'Zone 3␣
↪Power Consumption']
weather_cols = ['Temperature', 'Humidity', 'Wind Speed', 'general diffuse␣
↪flows', 'diffuse flows']
```

```
Dataset information after preprocessing:
<class 'pandas.core.frame.DataFrame'>
DatetimeIndex: 52416 entries, 2017-01-01 00:00:00 to 2017-12-30 23:50:00
Data columns (total 8 columns):
 #   Column                    Non-Null Count  Dtype
---  ------                    --------------  -----
 0   Temperature               52416 non-null  float64
 1   Humidity                  52416 non-null  float64
 2   Wind Speed                52416 non-null  float64
 3   general diffuse flows     52416 non-null  float64
 4   diffuse flows             52416 non-null  float64
 5   Zone 1 Power Consumption  52416 non-null  float64
 6   Zone 2 Power Consumption  52416 non-null  float64
 7   Zone 3 Power Consumption  52416 non-null  float64
dtypes: float64(8)
memory usage: 3.6 MB
```

[22]:
```
print("\nSample data:")
df.head()
```

Sample data:

[22]:

DateTime	Temperature	Humidity	Wind Speed	general diffuse flows \
2017-01-01 00:00:00	6.559	73.8	0.083	0.051
2017-01-01 00:10:00	6.414	74.5	0.083	0.070
2017-01-01 00:20:00	6.313	74.5	0.080	0.062
2017-01-01 00:30:00	6.121	75.0	0.083	0.091
2017-01-01 00:40:00	5.921	75.7	0.081	0.048

```
                     diffuse flows  Zone 1 Power Consumption  \
DateTime
2017-01-01 00:00:00          0.119               34055.69620
2017-01-01 00:10:00          0.085               29814.68354
2017-01-01 00:20:00          0.100               29128.10127
2017-01-01 00:30:00          0.096               28228.86076
2017-01-01 00:40:00          0.085               27335.69620

                     Zone 2 Power Consumption  Zone 3 Power Consumption
DateTime
2017-01-01 00:00:00               16128.87538               20240.96386
2017-01-01 00:10:00               19375.07599               20131.08434
2017-01-01 00:20:00               19006.68693               19668.43373
2017-01-01 00:30:00               18361.09422               18899.27711
2017-01-01 00:40:00               17872.34043               18442.40964
```

Exploratory Data Analysis (EDA)

Initial Exploration and Data Coverage

[23]:
```
# Check time coverage and frequency
print("\nDate range:", df.index.min(), "to", df.index.max())
time_diff = (df.index[1] - df.index[0]).seconds / 60
print("Sampling frequency:", time_diff, "minutes")
print("Number of observations:", len(df))
print("Expected observations for complete year at 10-min intervals:", 365*24*6)

# Summary statistics
print("\nSummary statistics:")
df.describe().T
```

```
Date range: 2017-01-01 00:00:00 to 2017-12-30 23:50:00
Sampling frequency: 10.0 minutes
Number of observations: 52416
Expected observations for complete year at 10-min intervals: 52560
Summary statistics:
```

[23]:
```
                            count          mean          std           min  \
Temperature               52416.0     18.810024     5.815476      3.247000
Humidity                  52416.0     68.259518    15.551177     11.340000
Wind Speed                52416.0      1.959489     2.348862      0.050000
general diffuse flows     52416.0    182.696614   264.400960      0.004000
diffuse flows             52416.0     75.028022   124.210949      0.011000
Zone 1 Power Consumption  52416.0  32344.970564  7130.562564  13895.696200
Zone 2 Power Consumption  52416.0  21042.509082  5201.465892   8560.081466
Zone 3 Power Consumption  52416.0  17835.406218  6622.165099   5935.174070
```

```
                                    25%           50%           75%  \
Temperature                   14.410000     18.780000     22.890000
Humidity                      58.310000     69.860000     81.400000
Wind Speed                     0.078000      0.086000      4.915000
general diffuse flows          0.062000      5.035500    319.600000
diffuse flows                  0.122000      4.456000    101.000000
Zone 1 Power Consumption   26310.668692  32265.920340  37309.018185
Zone 2 Power Consumption   16980.766032  20823.168405  24713.717520
Zone 3 Power Consumption   13129.326630  16415.117470  21624.100420

                                  max
Temperature                  40.01000
Humidity                     94.80000
Wind Speed                    6.48300
general diffuse flows      1163.00000
diffuse flows               936.00000
Zone 1 Power Consumption   52204.39512
Zone 2 Power Consumption   37408.86076
Zone 3 Power Consumption   47598.32636
```

Note:
The initial calculation (`365 × 24 × 6 = 52,560`) suggested that 144 observations were missing (`52,560 - 52,416 = 144`). It's important to verify such discrepancies to rule out systematic issues — for example, if a sensor had defaulted for an extended period, it could introduce bias into the model.

In our case, a closer inspection reveals that the data spans only 364 days (from January 1 to December 30, 2017), not the full 365. Therefore, the 144 "missing" records are simply due to one less day of data collection, and not a data quality issue.

Distribution of Features

[24]:
```python
# Plot histogram and KDE for numeric columns
num_cols = df.select_dtypes(include=np.number).columns
fig, axes = plt.subplots(3, 3, figsize=(18, 15))
axes = axes.flatten()

for i, col in enumerate(num_cols):
    if i < len(axes):
        sns.histplot(df[col].dropna(), kde=True, bins=40, ax=axes[i])
        axes[i].set_title(f"Distribution of {col}", fontsize=12)
        axes[i].set_xlabel("")
        mean_val = df[col].mean()
        median_val = df[col].median()
        axes[i].axvline(mean_val, color='r', linestyle='--', label=f'Mean:␣
↪{mean_val:.2f}')
```

```
        axes[i].axvline(median_val, color='g', linestyle='-.', label=f'Median:␣
 ↪{median_val:.2f}')
        axes[i].legend(fontsize=9)

for j in range(i+1, len(axes)):
    axes[j].set_visible(False)

plt.tight_layout()
plt.suptitle("Feature Distributions with Summary Statistics", y=1.02,␣
 ↪fontsize=16)
plt.show()
```

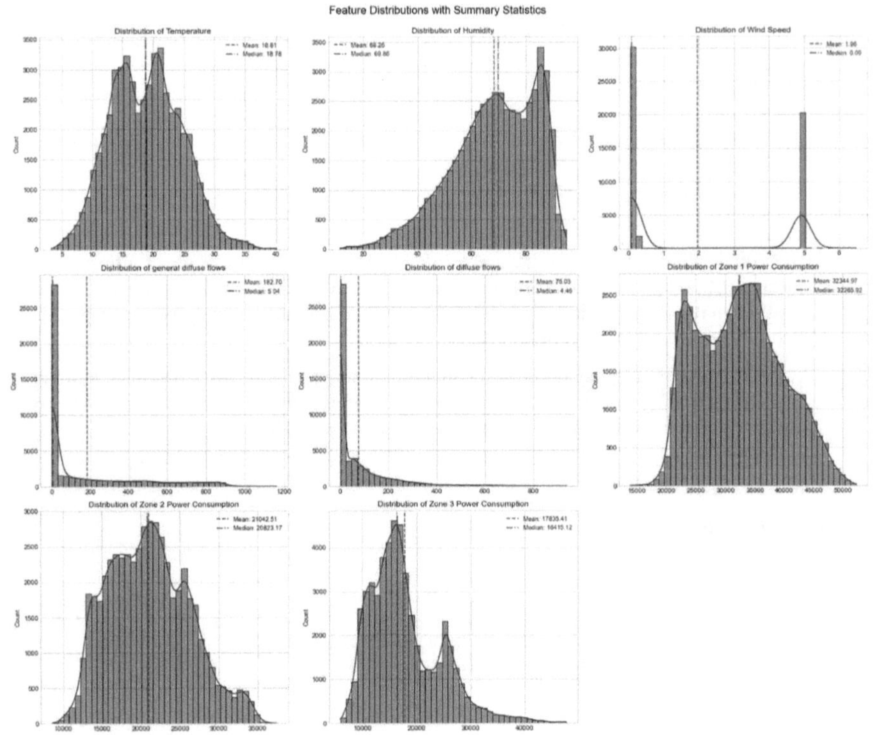

Outlier Detection and Its Importance in Energy Forecasting

- Outlier detection is a critical step in any machine learning pipeline, especially when working with real-world sensor or operational data like power consumption. Outliers can:
- Skew the distribution of data, affecting statistical summaries like mean and variance.

- Mislead model training by introducing noise or bias.
- Indicate genuine anomalies such as equipment malfunctions, spikes in demand, or sensor errors all of which may require further investigation or treatment.

[25]:
```
# Create boxplots to identify outliers
plt.figure(figsize=(16, 10))
for i, col in enumerate(power_cols):
    plt.subplot(3, 1, i+1)
    sns.boxplot(x=df[col])
    plt.title(f"Boxplot of {col}", fontsize=12)
    plt.tight_layout()
plt.suptitle("Outlier Detection Using Boxplots for Power Consumption", y=1.02,␣
 ↪fontsize=16)
plt.show()
```

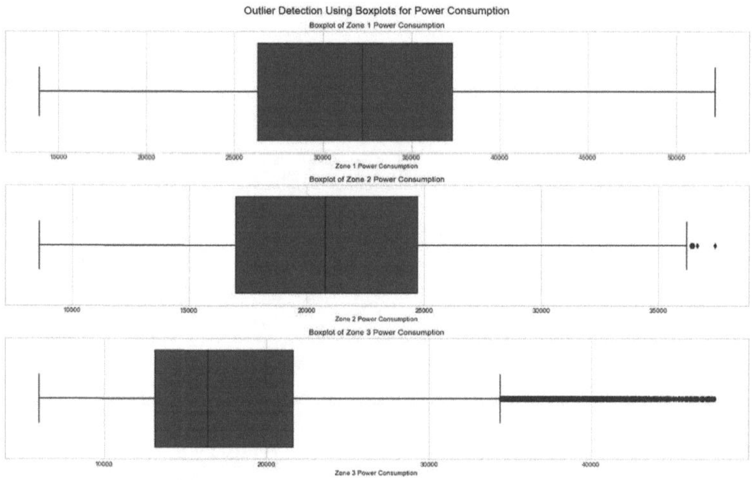

In the above boxplots: Zone 3 displays a **significant number of high-end outliers**, suggesting frequent power consumption surges/irregular demand patterns, or possible data logging issues. Zones 1 and 2 have relatively stable distributions with minimal extreme values.

These observations influence how we treat the data prior to modeling—for example, applying outlier clipping, transformation, or using robust models less sensitive to extreme values. In our case, they also highlight the potential need for zone-specific modeling strategies due to differing data characteristics.

Correlation Analysis

```
[26]: # Compute correlation matrix
corr_matrix = df.corr()

# Create heatmap of correlations
plt.figure(figsize=(12, 10))
mask = np.triu(corr_matrix)
sns.heatmap(corr_matrix, annot=True, fmt=".2f", cmap="coolwarm", mask=mask,
            vmin=-1, vmax=1, center=0, square=True, linewidths=.5)
plt.title("Correlation Matrix of Features", fontsize=15)
plt.tight_layout()
plt.show()
```

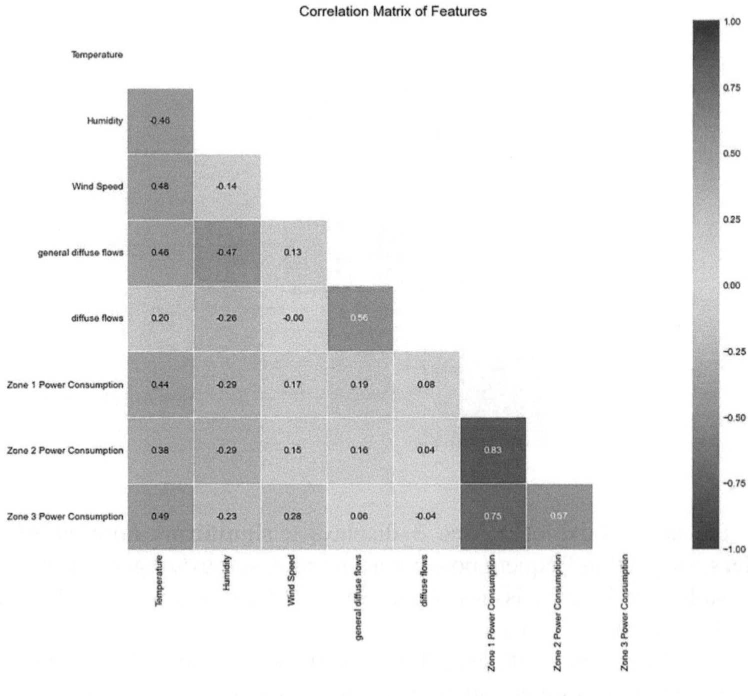

Correlation Matrix Analysis

Generating a correlation matrix is a common practice that provides valuable insights into how different features correlate with one another and with the target variables (zone-wise power consumption). Key takeaways from the above heatmap:

- **Strong inter-zone correlations**: Zone 1 and Zone 2 power consumption have a very high positive correlation (**0.83**), indicating similar usage patterns or shared influencing factors like grid structure or geographic proximity.

- **Temperature** shows positive correlation with all three zones (e.g., **0.44** with Zone 1), suggesting that electricity demand rises with warmer weather.
- **Humidity** has a mild negative correlation with power consumption (e.g., **−0.29** with Zones 1 and 2), possibly because humid days coincide with cooler temperatures in Tetouan. The strong negative correlation between **humidity and temperature (-0.46)** further supports this.
- **Diffuse solar radiation flows** (both general and specific) show weak or inconsistent correlations with power consumption. This implies they don't have an obvious linear relationship with electricity usage and may not be useful for **simple linear regression-style models**. However, more complex models (e.g., tree-based or neural networks) could still capture any nonlinear patterns they might contain.

These relationships can inform feature selection, highlight potential redundancy (e.g., between zones), and point to causal patterns worth investigating further with domain knowledge.

[27]:
```
# Focusing on correlations with the target variable (Power consumption)
power_corr = corr_matrix[power_cols].drop(power_cols)
plt.figure(figsize=(12, 8))
sns.heatmap(power_corr, annot=True, fmt=".2f", cmap="coolwarm",
            vmin=-1, vmax=1, center=0, linewidths=.5)
plt.title("Correlation of Features with Power Consumption", fontsize=15)
plt.tight_layout()
plt.show()
```

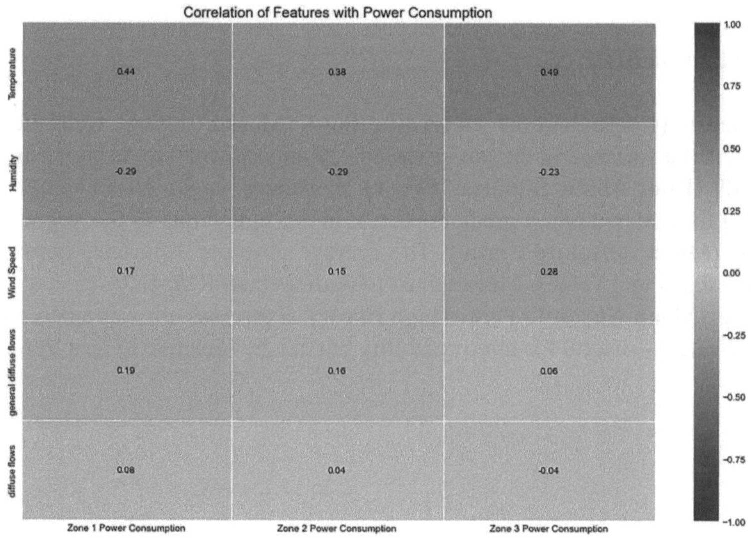

Model Training and Evaluation

[28]:
```
features = ['Temperature', 'Humidity', 'Wind Speed', 'general diffuse flows',
        'diffuse flows']
target = 'Zone 1 Power Consumption'
```

Model Evaluation: Metrics and Validation Strategy

To assess how well our models learn from data and generalize to unseen scenarios, we split our dataset into **training** and **testing** sets. The training set is used to fit the model, while the test set helps us evaluate how well the model performs on new, unseen data.

Why Compare Train vs. Test Performance?

- **Training error** shows how well the model fits the known data.
- **Testing error** reveals how well the model actually performed in an unseen scenario/production.
- A large gap between the two often indicates **overfitting** (i.e model memorizes training data but fails on new data).

Key Evaluation Metrics

- **R^2 Score (Coefficient of Determination)**: Measures how well the model explains the variance in the target variable. Ranges from 0 to 1 (higher is better).
- **RMSE (Root Mean Squared Error)**: Penalizes larger errors more heavily. Useful for understanding prediction error in the same units as the target.
- **MAE (Mean Absolute Error)**: The average absolute difference between predicted and actual values. More robust to outliers than RMSE.
- **MAPE (Mean Absolute Percentage Error)**: Expresses error as a percentage of actual values—useful for interpretability but can be sensitive to near-zero values.

By comparing these metrics across models, we can choose the best-performing one and identify whether further tuning or feature engineering is needed.

```
[29]: # Helper function to evaluate models
def evaluate_model(name, model, X_train, y_train, X_test, y_test):
    model.fit(X_train, y_train)
    y_pred_train = model.predict(X_train)
    y_pred_test = model.predict(X_test)

    metrics = {
        'Model': name,
        'Train R2': r2_score(y_train, y_pred_train),
        'Test R2': r2_score(y_test, y_pred_test),
        'Train RMSE': np.sqrt(mean_squared_error(y_train, y_pred_train)),
        'Test RMSE': np.sqrt(mean_squared_error(y_test, y_pred_test)),
        'Train MAE': mean_absolute_error(y_train, y_pred_train),
        'Test MAE': mean_absolute_error(y_test, y_pred_test),
        'Test MAPE': np.mean(np.abs((y_test - y_pred_test) / (y_test + 1e-5))) *␣
↪100
    }
    return metrics

# Define models to compare
models = {
    "Linear Regression": LinearRegression(),
    "Ridge Regression": Ridge(alpha=1.0),
    "Lasso Regression": Lasso(alpha=0.01),
    "Random Forest": RandomForestRegressor(n_estimators=100, random_state=42),
    "Gradient Boosting": GradientBoostingRegressor(n_estimators=100,␣
↪random_state=42)
}

# Prepare data
X = df[features]
y = df[target]
split_index = int(len(df) * 0.8)
X_train, X_test = X.iloc[:split_index], X.iloc[split_index:]
```

```
y_train, y_test = y.iloc[:split_index], y.iloc[split_index:]
```

[30]:
```
# Evaluate each model
results = []
for name, model in models.items():
    metrics = evaluate_model(name, model, X_train, y_train, X_test, y_test)
    results.append(metrics)

results_df = pd.DataFrame(results)
results_df.set_index("Model", inplace=True)
results_df.sort_values("Test R2", ascending=False, inplace=True)

# Show results
display(results_df)
```

```
                   Train R2   Test R2   Train RMSE    Test RMSE    Train MAE  \
Model
Gradient Boosting  0.362081  0.048759  5729.333537  6014.169853  4507.037797
Lasso Regression   0.192089  0.044084  6447.676267  6028.931950  5314.320299
Ridge Regression   0.192089  0.044084  6447.676267  6028.932342  5314.320325
Linear Regression  0.192089  0.044084  6447.676267  6028.932599  5314.320342
Random Forest      0.947564 -0.223839  1642.615408  6821.700465  1089.681304

                      Test MAE  Test MAPE
Model
Gradient Boosting  4774.936183  17.942949
Lasso Regression   5082.744315  18.928821
Ridge Regression   5082.744596  18.928823
Linear Regression  5082.744456  18.928821
Random Forest      5008.637028  18.873822
```

[31]:
```
results_df.to_csv('basic_nb_eval_results.csv')
```

Conclusion

The results above provide a comparative view of how different models perform on the energy consumption forecasting task:

- **Gradient Boosting** delivers the best overall performance, achieving the highest test R^2 (0.0488) and the lowest test RMSE (~6014). While the R^2 values are modest, this model demonstrates some ability to generalize beyond the training data.
- **MAPE** values for all models are in the range of 17–19%, with Gradient Boosting achieving the lowest at approximately 17.94%, reinforcing its relative advantage.
- **Linear Regression**, **Ridge**, and **Lasso Regression** show nearly identical results, indicating that the current feature set does not strongly support linear relationships. Their low test R^2 scores (~0.044) suggest that the raw features are insufficient for accurate prediction with simple linear models.
- **Random Forest** shows clear signs of **overfitting**: it has an extremely high training R^2 (**0.95**) but a negative test R^2 (**–0.22**), which means it fails to generalize and performs even worse than predicting the mean. This highlights

that using a more complex or powerful machine learning model **does not inherently lead to better performance**—especially when the underlying data lacks informative features or is not well-processed.

These findings underscore the need for stronger feature engineering and domain knowledge-informed transformations to improve predictive performance—a key focus of the advanced notebook that follows.

Power Consumption Forecasting for Tetouan City: Advanced

This section is intended for readers who are familiar with machine learning and want to explore how applying their **domain knowledge** in **feature engineering** can significantly boost model performance—even for simple models like linear regression.

We'll demonstrate how crafting **derived features** from temporal patterns, weather variables, and energy usage insights can uncover hidden structure in the data. By bridging data science skillsets with renewable energy knowledge, we demonstrate that thoughtful feature design can outperform complex models that lack domain alignment.

Data Loading and Initial Setup

[13]:
```
#! pip install -r requirements.txt
```

[14]:
```
import pandas as pd
import numpy as np
import matplotlib.pyplot as plt
import seaborn as sns
import warnings
warnings.filterwarnings('ignore')
import requests
import zipfile
import io
from sklearn.linear_model import LinearRegression
from sklearn.ensemble import RandomForestRegressor
from sklearn.metrics import mean_absolute_error, mean_squared_error, r2_score
import matplotlib.dates as mdates
import plotly.express as px
from statsmodels.tsa.seasonal import seasonal_decompose

# Set random seed for reproducibility
np.random.seed(42)

# Set better plot aesthetics
plt.style.use('seaborn-whitegrid')
sns.set_palette("viridis")
```

[15]:
```
# Download and extract CSV from ZIP
url = "https://archive.ics.uci.edu/static/public/849/
 ↪power+consumption+of+tetouan+city.zip"
response = requests.get(url)
with zipfile.ZipFile(io.BytesIO(response.content)) as z:
    with z.open(z.namelist()[0]) as f:
        df = pd.read_csv(f)

# Display the first few rows
print("Dataset initial shape:", df.shape)
df.head()
```

```
Dataset initial shape: (52416, 9)
```

[15]:
```
        DateTime  Temperature  Humidity  Wind Speed  general diffuse flows  \
0  1/1/2017 0:00        6.559      73.8       0.083                  0.051
1  1/1/2017 0:10        6.414      74.5       0.083                  0.070
2  1/1/2017 0:20        6.313      74.5       0.080                  0.062
3  1/1/2017 0:30        6.121      75.0       0.083                  0.091
4  1/1/2017 0:40        5.921      75.7       0.081                  0.048

   diffuse flows  Zone 1 Power Consumption  Zone 2  Power Consumption  \
0          0.119               34055.69620                16128.87538
1          0.085               29814.68354                19375.07599
2          0.100               29128.10127                19006.68693
3          0.096               28228.86076                18361.09422
4          0.085               27335.69620                17872.34043

   Zone 3  Power Consumption
0                20240.96386
1                20131.08434
2                19668.43373
3                18899.27711
4                18442.40964
```

[16]:
```
# Removes extra spacing from column names for consistency
df.rename(columns={
    'Zone 2  Power Consumption': 'Zone 2 Power Consumption',
    'Zone 3  Power Consumption': 'Zone 3 Power Consumption'
}, inplace=True)

# Convert 'DateTime' column which is currently an 'object' datatype to␣
 ↪'datetime' datatype and set it as index
df['DateTime'] = pd.to_datetime(df['DateTime'], dayfirst=True)
df.set_index('DateTime', inplace=True)

# Preview structure
```

```
print("\nDataset information after preprocessing:")
df.info()
# Define key column groups for easier reference
power_cols = ['Zone 1 Power Consumption', 'Zone 2 Power Consumption', 'Zone 3␣
↪Power Consumption']
weather_cols = ['Temperature', 'Humidity', 'Wind Speed', 'general diffuse␣
↪flows', 'diffuse flows']
```

```
Dataset information after preprocessing:
<class 'pandas.core.frame.DataFrame'>
DatetimeIndex: 52416 entries, 2017-01-01 00:00:00 to 2017-12-30 23:50:00
Data columns (total 8 columns):
 #   Column                    Non-Null Count  Dtype
---  ------                    --------------  -----
 0   Temperature               52416 non-null  float64
 1   Humidity                  52416 non-null  float64
 2   Wind Speed                52416 non-null  float64
 3   general diffuse flows     52416 non-null  float64
 4   diffuse flows             52416 non-null  float64
 5   Zone 1 Power Consumption  52416 non-null  float64
 6   Zone 2 Power Consumption  52416 non-null  float64
 7   Zone 3 Power Consumption  52416 non-null  float64
dtypes: float64(8)
memory usage: 3.6 MB
```

[17]:
```
print("\nSample data:")
df.head()
```

```
Sample data:
```

[17]:
```
                     Temperature  Humidity  Wind Speed  general diffuse flows  \
DateTime
2017-01-01 00:00:00        6.559      73.8       0.083                  0.051
2017-01-01 00:10:00        6.414      74.5       0.083                  0.070
2017-01-01 00:20:00        6.313      74.5       0.080                  0.062
2017-01-01 00:30:00        6.121      75.0       0.083                  0.091
2017-01-01 00:40:00        5.921      75.7       0.081                  0.048

                     diffuse flows  Zone 1 Power Consumption  \
DateTime
2017-01-01 00:00:00          0.119               34055.69620
2017-01-01 00:10:00          0.085               29814.68354
2017-01-01 00:20:00          0.100               29128.10127
2017-01-01 00:30:00          0.096               28228.86076
2017-01-01 00:40:00          0.085               27335.69620

                     Zone 2 Power Consumption  Zone 3 Power Consumption
DateTime
2017-01-01 00:00:00               16128.87538               20240.96386
2017-01-01 00:10:00               19375.07599               20131.08434
2017-01-01 00:20:00               19006.68693               19668.43373
2017-01-01 00:30:00               18361.09422               18899.27711
2017-01-01 00:40:00               17872.34043               18442.40964
```

Advanced Data Pre-processing

Missing Values

Simulating Missing Data for Demonstration

The original dataset does not contain any missing values, which is great for modeling but limits our ability to demonstrate how to detect and handle missing data—a common challenge in real-world energy datasets.

- To illustrate missing data analysis techniques, we create a synthetic version of the dataset by randomly introducing missing values across all columns (about 1% per column). Additionally, we simulate a contiguous block of missing data in the "Zone 1 Power Consumption" column to mimic a sensor outage or data logging failure.
- This synthetic dataset allows us to visualize missingness patterns, calculate missing data percentages, and identify the longest contiguous gaps—all essential steps in preparing time series data for machine learning.

```
[18]: # Step 1: Generate synthetic data
synthetic_df = df.copy()
np.random.seed(42)
for col in synthetic_df.columns:
    missing_indices = np.random.choice(synthetic_df.index, size=int(0.01 *␣
 ↪len(synthetic_df)), replace=False)
    synthetic_df.loc[missing_indices, col] = np.nan
synthetic_df.iloc[10000:10030, synthetic_df.columns.get_loc('Zone 1 Power␣
 ↪Consumption')] = np.nan

# Step 2: Visualize missing data
missing_percentage = (synthetic_df.isna().sum() / len(synthetic_df)) * 100
missing_df = pd.DataFrame({'Column': synthetic_df.columns, 'Missing (%)':␣
 ↪missing_percentage.values})

plt.figure(figsize=(10, 5))
sns.barplot(x='Missing (%)', y='Column', data=missing_df.sort_values('Missing␣
 ↪(%)', ascending=False), palette="Blues_d")
plt.title("Percentage of Missing Data by Column", fontsize=14)
plt.xlabel("Missing Data (%)")
plt.ylabel("Feature")
plt.xlim(0, missing_percentage.max() + 1)
```

```
plt.grid(axis='x', linestyle='--', alpha=0.6)
plt.tight_layout()
plt.show()

# Step 3: Contiguous gap detection
missing_periods = {}
for col in synthetic_df.columns:
    is_na = synthetic_df[col].isna()
    if not is_na.any():
        continue
    na_groups = (is_na != is_na.shift()).cumsum()
    runs = na_groups[is_na].value_counts().sort_index()
    longest = runs.max()
    missing_periods[col] = longest

print("\nLongest contiguous periods of missing data (in 10-minute intervals):")
for col, length in missing_periods.items():
    print(f"{col}: {length} intervals ({length * 10 / 60:.1f} hours)")
```

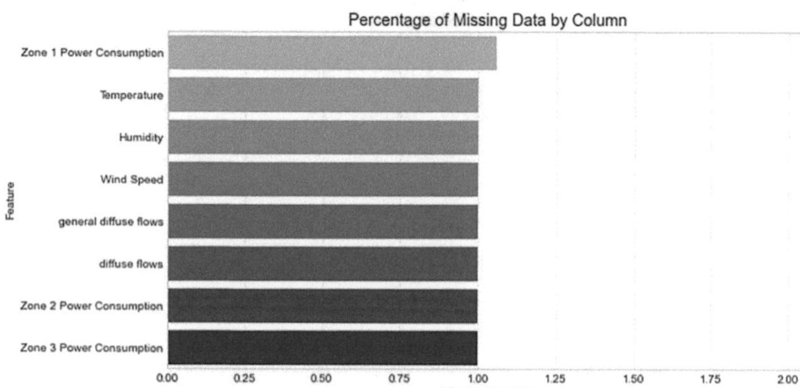


```
Longest contiguous periods of missing data (in 10-minute intervals):
Temperature: 2 intervals (0.3 hours)
Humidity: 2 intervals (0.3 hours)
Wind Speed: 2 intervals (0.3 hours)
general diffuse flows: 2 intervals (0.3 hours)
diffuse flows: 2 intervals (0.3 hours)
Zone 1 Power Consumption: 30 intervals (5.0 hours)
Zone 2 Power Consumption: 2 intervals (0.3 hours)
Zone 3 Power Consumption: 2 intervals (0.3 hours)
```

[19]:
```
# Interpolate weather-related features with appropriate method
print("Missing values before interpolation:", synthetic_df[weather_cols].isna().
    ↪sum())
synthetic_df[weather_cols] = synthetic_df[weather_cols].
    ↪interpolate(method='time')

# For power consumption, use a combination of interpolation methods
synthetic_df[power_cols] = synthetic_df[power_cols].interpolate(method='linear',␣
    ↪limit=6)
synthetic_df[power_cols] = synthetic_df[power_cols].interpolate(method='time')
synthetic_df[power_cols] = synthetic_df[power_cols].fillna(method='ffill').
    ↪fillna(method='bfill')

# Verify missing value treatment
print("Missing values after interpolation:", synthetic_df.isna().sum())

remaining_na = synthetic_df.isna().sum().sum()
if remaining_na > 0:
    print(f"Removing {synthetic_df.isna().any(axis=1).sum()} rows with remaining␣
    ↪NaN values")
    synthetic_df = synthetic_df.dropna()
else:
    print("No NaN values remaining in the dataset")
```

```
Missing values before interpolation: Temperature              524
Humidity                 524
Wind Speed               524
general diffuse flows    524
diffuse flows            524
dtype: int64
Missing values after interpolation: Temperature                0
Humidity                    0
Wind Speed                  0
general diffuse flows       0
diffuse flows               0
Zone 1 Power Consumption    0
Zone 2 Power Consumption    0
Zone 3 Power Consumption    0
dtype: int64
No NaN values remaining in the dataset
```

Outlier Detection and Treatment

[20]:
```
# Define outlier detection and capping function
def detect_and_cap_outliers(df, column, lower_quantile=0.01, upper_quantile=0.
↪99, plot=True):
    original_values = df[column].copy()
    lower_bound = df[column].quantile(lower_quantile)
    upper_bound = df[column].quantile(upper_quantile)
    outliers = (df[column] < lower_bound) | (df[column] > upper_bound)
    outlier_count = outliers.sum()
    outlier_percentage = (outlier_count / len(df)) * 100

    print(f"{column} - Outliers: {outlier_count} ({outlier_percentage:.2f}%)")
    print(f"  Lower bound: {lower_bound:.2f}, Upper bound: {upper_bound:.2f}")

    df[column] = df[column].clip(lower=lower_bound, upper=upper_bound)

    if plot and outlier_count > 0:
        plt.figure(figsize=(14, 6))
        plt.subplot(1, 2, 1)
        sns.histplot(original_values, bins=50, kde=True)
        plt.axvline(lower_bound, color='r', linestyle='--')
        plt.axvline(upper_bound, color='r', linestyle='--')
        plt.title(f"Before Outlier Treatment: {column}")
        plt.subplot(1, 2, 2)
        sns.histplot(df[column], bins=50, kde=True)
        plt.axvline(lower_bound, color='g', linestyle='--')
        plt.axvline(upper_bound, color='g', linestyle='--')
        plt.title(f"After Outlier Treatment: {column}")
        plt.tight_layout()
        plt.show()

    return df

# Apply to power and weather columns
print("Detecting and treating outliers in power consumption data:")
for col in power_cols:
    df = detect_and_cap_outliers(df, col)

print("\nDetecting and treating outliers in weather data:")
for col in weather_cols:
    df = detect_and_cap_outliers(df, col, plot=False)
```

```
Detecting and treating outliers in power consumption data:
Zone 1 Power Consumption - Outliers: 1050 (2.00%)
  Lower bound: 20274.81, Upper bound: 48099.48
```

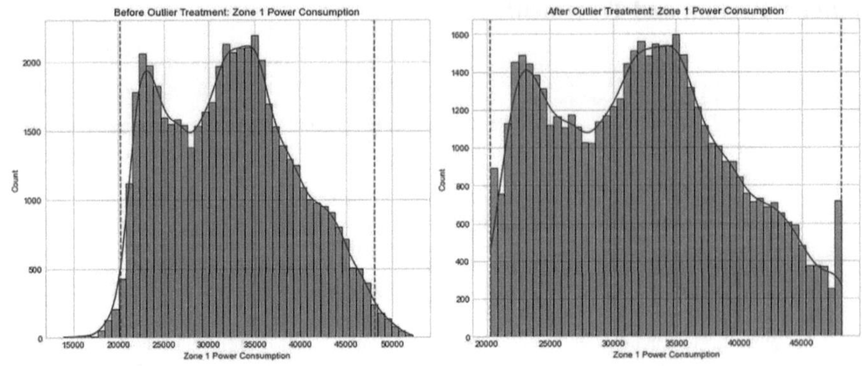


```
Zone 2 Power Consumption - Outliers: 1050 (2.00%)
  Lower bound: 11631.70, Upper bound: 33734.95
```

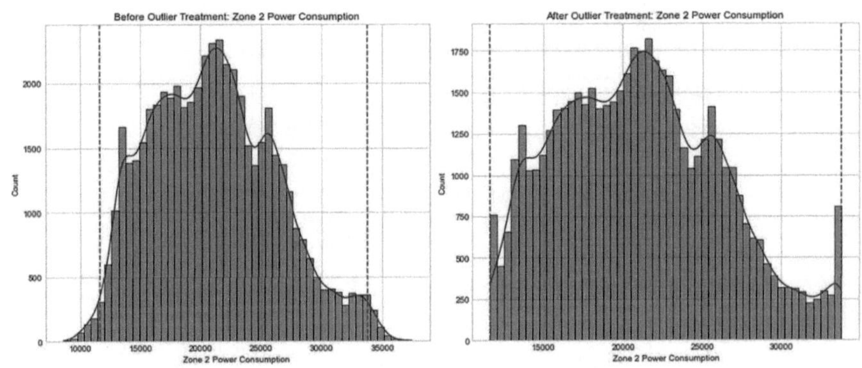


```
Zone 3 Power Consumption - Outliers: 1048 (2.00%)
  Lower bound: 7726.27, Upper bound: 38722.44
```

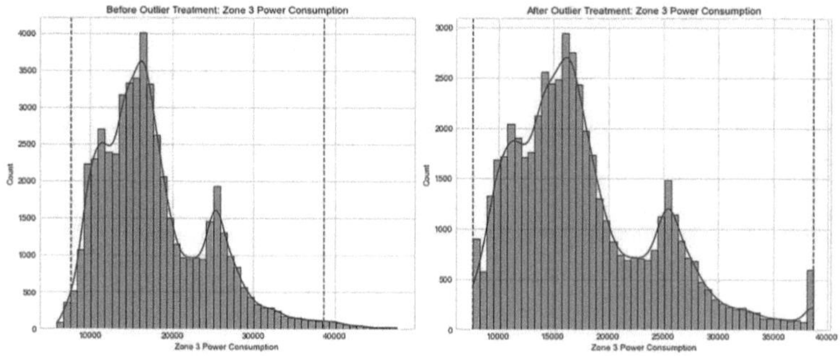


```
Detecting and treating outliers in weather data:
Temperature - Outliers: 1050 (2.00%)
  Lower bound: 6.76, Upper bound: 33.16
Humidity - Outliers: 1041 (1.99%)
  Lower bound: 27.33, Upper bound: 92.00
Wind Speed - Outliers: 1030 (1.97%)
  Lower bound: 0.07, Upper bound: 4.93
general diffuse flows - Outliers: 846 (1.61%)
  Lower bound: 0.02, Upper bound: 885.00
diffuse flows - Outliers: 934 (1.78%)
  Lower bound: 0.06, Upper bound: 596.80
```

Advanced Feature Engineering

Temporal Patterns in Power Consumption

[21]:
```
# Add temporal features for grouping
df['hour'] = df.index.hour
df['day'] = df.index.day
df['weekday'] = df.index.weekday
df['month'] = df.index.month
df['quarter'] = df.index.quarter
df['is_weekend'] = df['weekday'].isin([5, 6]).astype(int)

# Resample to daily data for easier visualization of trends
df_daily = df[power_cols].resample('D').mean()
```

[22]:
```
# Plot daily average power consumption over time
plt.figure(figsize=(16, 8))
```

```
for col in power_cols:
    plt.plot(df_daily.index, df_daily[col], label=col)
plt.title("Daily Average Power Consumption Over Time", fontsize=15)
plt.xlabel("Date")
plt.ylabel("Power Consumption (kW)")
plt.legend()
plt.grid(True)
plt.gca().xaxis.set_major_formatter(mdates.DateFormatter('%b %Y'))
plt.gca().xaxis.set_major_locator(mdates.MonthLocator(interval=1))
plt.xticks(rotation=45)
plt.tight_layout()
plt.show()
```

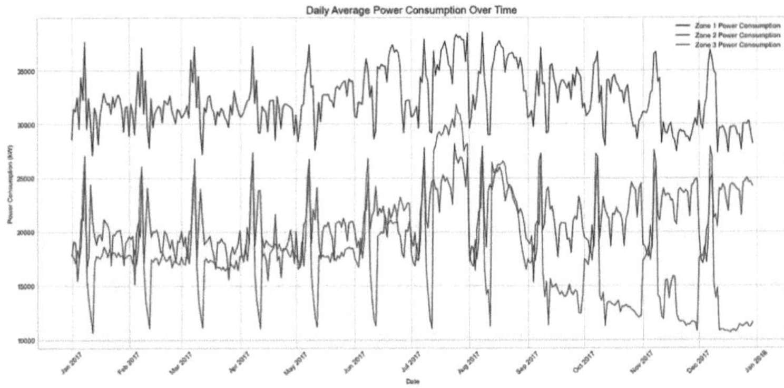

[23]:

```
# Plot hourly patterns
hourly_avg = df.groupby('hour')[power_cols].mean()
plt.figure(figsize=(14, 7))
for col in power_cols:
    plt.plot(hourly_avg.index, hourly_avg[col], marker='o', linewidth=2,␣
 ↪label=col)
plt.title("Average Hourly Power Consumption Pattern", fontsize=15)
plt.xlabel("Hour of Day")
plt.ylabel("Power Consumption (kW)")
plt.xticks(range(0, 24))
plt.grid(True)
plt.legend()
plt.tight_layout()
plt.show()
```

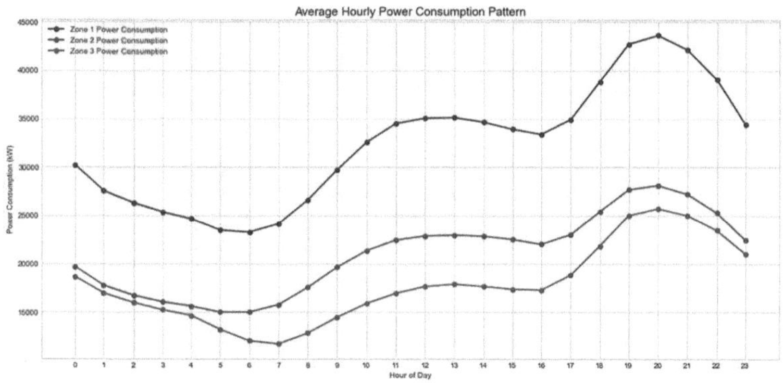

[24]:
```
# Weekly pattern with standard deviation
weekly_avg = df.groupby('weekday')[power_cols].agg(['mean', 'std'])
days = ['Monday', 'Tuesday', 'Wednesday', 'Thursday', 'Friday', 'Saturday',␣
 ↪'Sunday']

plt.figure(figsize=(14, 8))
for i, col in enumerate(power_cols):
    plt.subplot(3, 1, i+1)
    means = weekly_avg[col]['mean'].values
    stds = weekly_avg[col]['std'].values
    plt.bar(days, means, yerr=stds, capsize=10, alpha=0.7)
    plt.title(f"Weekly Pattern: {col}", fontsize=12)
    plt.ylabel("Power Consumption (kW)")
    plt.grid(axis='y', linestyle='--', alpha=0.7)
plt.tight_layout()
plt.suptitle("Power Consumption by Day of Week with Standard Deviation", y=1.02,␣
 ↪fontsize=15)
plt.show()
```

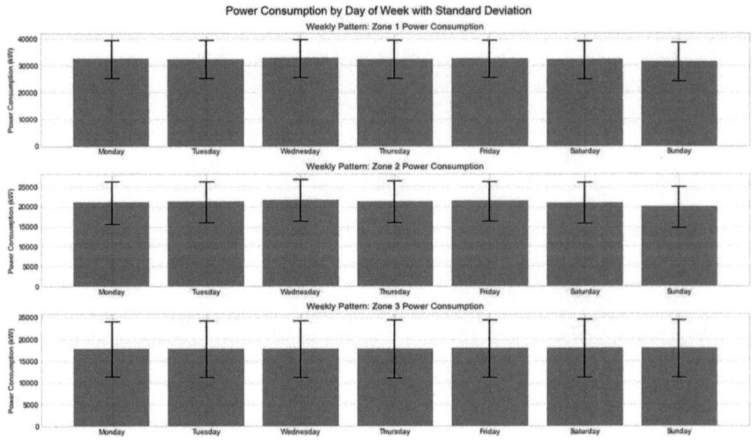

[25]:
```
# Monthly patterns
monthly_avg = df.groupby('month')[power_cols].mean()
months = ['Jan', 'Feb', 'Mar', 'Apr', 'May', 'Jun', 'Jul', 'Aug', 'Sep', 'Oct',␣
↪'Nov', 'Dec']

plt.figure(figsize=(14, 8))
for i, col in enumerate(power_cols):
    plt.subplot(3, 1, i+1)
    plt.plot(months, monthly_avg[col].values, marker='o', linewidth=2)
    plt.title(f"Monthly Pattern: {col}", fontsize=12)
    plt.ylabel("Power Consumption (kW)")
    plt.grid(True, alpha=0.3)
plt.tight_layout()
plt.suptitle("Seasonal Power Consumption Patterns", y=1.02, fontsize=15)
plt.show()
```

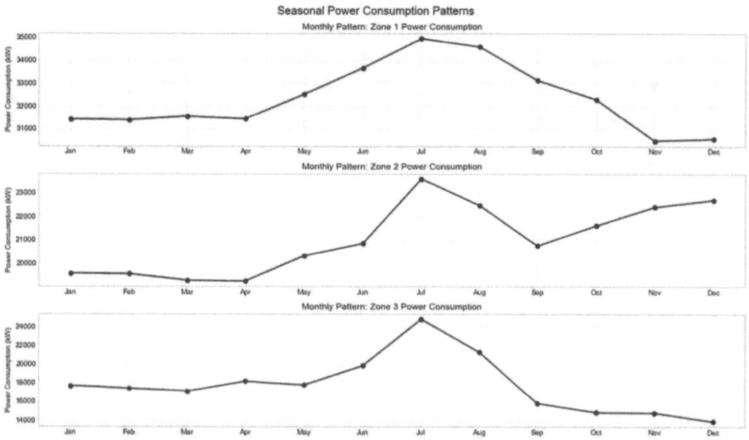

[26]:
```
# Time series decomposition
for col in power_cols:
    hourly_data = df[col].resample('H').mean()
    decomposition = seasonal_decompose(hourly_data.dropna(), period=24)
    fig, axes = plt.subplots(4, 1, figsize=(14, 12))
    decomposition.observed.plot(ax=axes[0], title='Observed')
    decomposition.trend.plot(ax=axes[1], title='Trend')
    decomposition.seasonal.plot(ax=axes[2], title='Seasonality')
    decomposition.resid.plot(ax=axes[3], title='Residuals')
    plt.tight_layout()
    plt.suptitle(f"Time Series Decomposition for {col}", y=1.02, fontsize=15)
    plt.show()
```

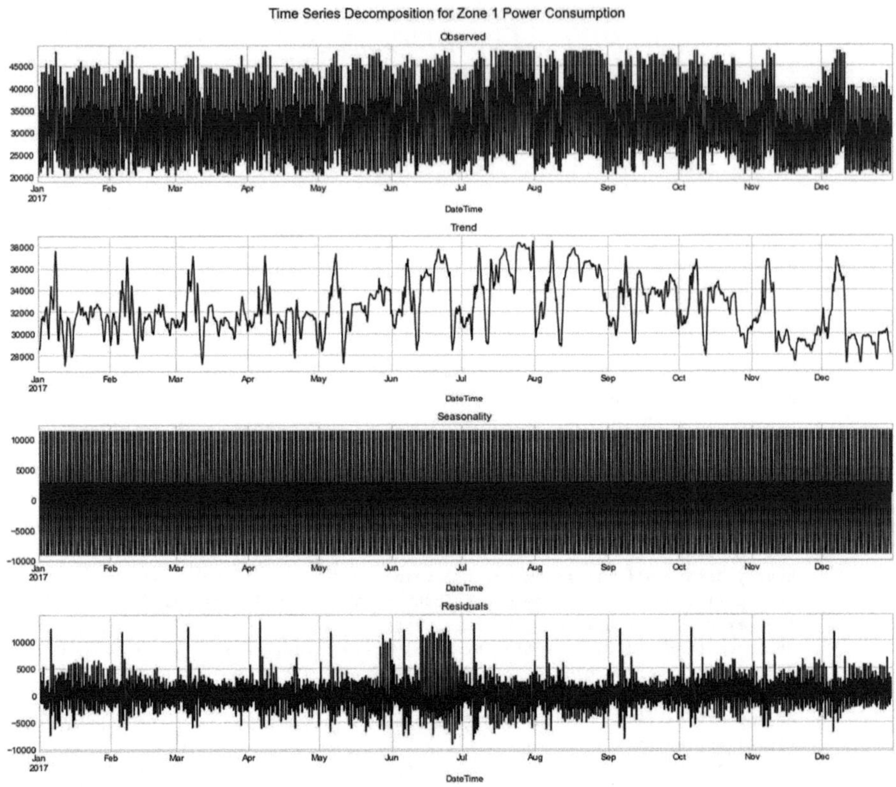
Time Series Decomposition for Zone 1 Power Consumption
Observed
45000
40000
35000
30000
25000
20000
Jan
2017
Feb
Mar
Apr
May
Jun
Jul
Aug
Sep
Oct
Nov
Dec
DateTime
Trend
38000
36000
34000
32000
30000
28000
DateTime
Seasonality
10000
5000
0
-5000
-10000
DateTime
Residuals
10000
5000
0
-5000
-10000
DateTime

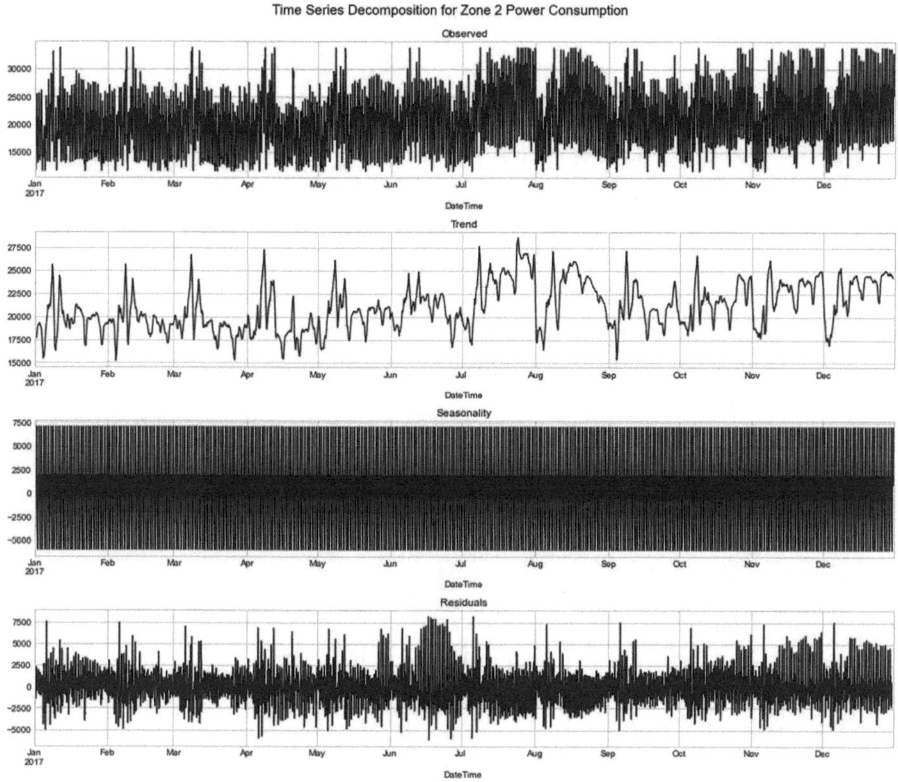
Time Series Decomposition for Zone 2 Power Consumption
Observed
Trend
Seasonality
Residuals
DateTime
Jan 2017
Feb
Mar
Apr
May
Jun
Jul
Aug
Sep
Oct
Nov
Dec

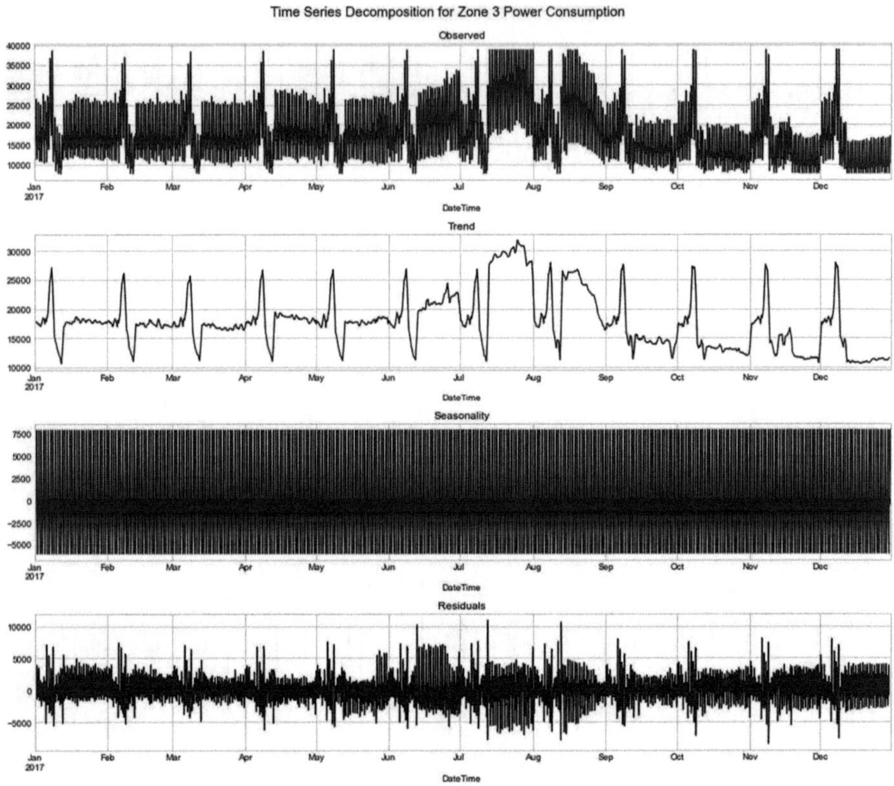

Weather Impact on Consumption

[27]:
```
# Create interactive scatter plots for weather vs power consumption
for zone in power_cols:
    fig = px.scatter(df.sample(5000),
                     x="Temperature",
                     y=zone,
                     color="hour",
                     hover_data=["Wind Speed", "Humidity"],
                     title=f"Temperature vs {zone} by Hour of Day",
                     labels={"Temperature": "Temperature (°C)", zone: "Power␣
↪Consumption (kW)"},
                     color_continuous_scale=px.colors.sequential.Viridis)
    fig.update_layout(width=900, height=600)
    fig.show()
```

[28]:
```
# Create a grid of scatter plots showing relationships between weather variables␣
↪and power consumption
fig, axes = plt.subplots(len(weather_cols), len(power_cols), figsize=(18, 15))

for i, weather in enumerate(weather_cols):
    for j, zone in enumerate(power_cols):
        sample_df = df.sample(5000)
        sns.regplot(x=weather, y=zone, data=sample_df, scatter_kws={'alpha':0.3},
                    line_kws={'color':'red'}, ax=axes[i, j])
        corr = df[weather].corr(df[zone])
        axes[i, j].annotate(f' = {corr:.2f}', xy=(0.05, 0.95), xycoords='axes␣
↪fraction')
        if i == len(weather_cols)-1:
            axes[i, j].set_xlabel(weather)
        else:
            axes[i, j].set_xlabel('')
        if j == 0:
            axes[i, j].set_ylabel(zone)
        else:
            axes[i, j].set_ylabel('')

plt.tight_layout()
plt.suptitle("Relationship Between Weather Variables and Power Consumption", y=1.
↪02, fontsize=16)
plt.show()
```

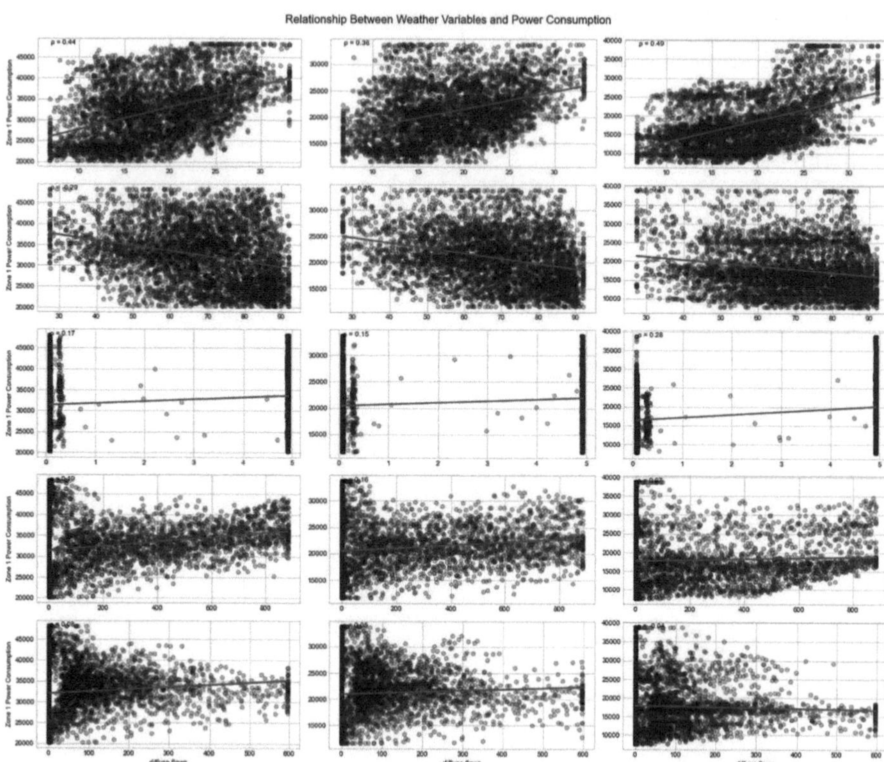

Enhanced Temporal and Cyclical Features

[29]:
```
# Temporal features
df['is_workday'] = (~df['weekday'].isin([5, 6])).astype(int)

# Time of day categories
df['time_of_day'] = pd.cut(
    df['hour'],
    bins=[0, 6, 12, 18, 24],
    labels=['Night', 'Morning', 'Afternoon', 'Evening'],
    include_lowest=True
)

# One-hot encode time of day
tod_dummies = pd.get_dummies(df['time_of_day'], prefix='tod')
df = pd.concat([df, tod_dummies], axis=1)
```

```
# Cyclical encoding
def cyclical_encoding(df, col, period):
    df[f'{col}_sin'] = np.sin(2 * np.pi * df[col] / period)
    df[f'{col}_cos'] = np.cos(2 * np.pi * df[col] / period)
    return df

df = cyclical_encoding(df, 'hour', 24)
df = cyclical_encoding(df, 'weekday', 7)
df = cyclical_encoding(df, 'month', 12)
df = cyclical_encoding(df, 'day', 31)

# Weekend-hour interaction
for h in range(24):
    df[f'weekend_hour_{h}'] = ((df['hour'] == h) &
                                (df['is_weekend'] == 1)).astype(int)
```

Advanced Lag and Rolling Features

[30]:
```
# Define lag intervals (10-minute intervals)
lag_intervals = {
    '30min': 3,
    '1hr': 6,
    '3hr': 18,
    '6hr': 36,
    '12hr': 72,
    '1day': 144,
    '1week': 1008
}

# Create lag features
for col in power_cols:
    for name, lag in lag_intervals.items():
        df[f'{col}_lag_{name}'] = df[col].shift(lag)
```

[31]:
```
# Create rolling statistics for power consumption
rolling_windows = [6, 12, 36, 72, 144]  # corresponds to 1hr, 2hr, 6hr, 12hr,␣
↪1day

for col in power_cols:
    for window in rolling_windows:
        df[f'{col}_roll_mean_{window}'] = df[col].rolling(window=window,␣
↪min_periods=1).mean()
        df[f'{col}_roll_std_{window}'] = df[col].rolling(window=window,␣
↪min_periods=1).std()
        df[f'{col}_roll_min_{window}'] = df[col].rolling(window=window,␣
↪min_periods=1).min()
        df[f'{col}_roll_max_{window}'] = df[col].rolling(window=window,␣
↪min_periods=1).max()
        df[f'{col}_roll_range_{window}'] = (
            df[f'{col}_roll_max_{window}'] - df[f'{col}_roll_min_{window}']
        )

# Rolling means for weather features
for col in weather_cols:
    df[f'{col}_roll_mean_6'] = df[col].rolling(window=6, min_periods=1).mean()
    df[f'{col}_roll_mean_36'] = df[col].rolling(window=36, min_periods=1).mean()
```

Derived and Interaction Features

[32]:
```
# Weather interaction features
df['temp_x_humidity'] = df['Temperature'] * df['Humidity'] / 100
df['temp_x_wind'] = df['Temperature'] * df['Wind Speed']
df['feels_like'] = df['Temperature'] - 0.2 * (100 - df['Humidity']) / 5
df['wind_chill'] = (
    13.12 + 0.6215 * df['Temperature'] -
    11.37 * (df['Wind Speed'] ** 0.16) +
    0.3965 * df['Temperature'] * (df['Wind Speed'] ** 0.16)
)

# Power consumption aggregates
df['total_power'] = df[power_cols].sum(axis=1)
df['avg_power'] = df[power_cols].mean(axis=1)

# Create zone ratios
for i in range(len(power_cols)):
    for j in range(i+1, len(power_cols)):
        col_i = power_cols[i]
        col_j = power_cols[j]
        ratio_name = f'ratio_{col_i.split()[1]}_{col_j.split()[1]}'
        df[ratio_name] = df[col_i] / (df[col_j] + 1)  # Avoid division by zero
```

[33]:
```
# Rate of change and percent change
for col in power_cols:
    df[f'{col}_diff'] = df[col].diff()
    df[f'{col}_pct_change'] = df[col].pct_change() * 100

# Time since peak feature
for col in power_cols:
    threshold = df[col].quantile(0.95)
    df[f'{col}_is_peak'] = (df[col] > threshold).astype(int)
    peak_indices = df.index[df[f'{col}_is_peak'] == 1]
    time_since_peak = pd.Series(index=df.index)
    for idx in df.index:
        previous_peaks = peak_indices[peak_indices < idx]
        if len(previous_peaks) > 0:
            most_recent_peak = previous_peaks[-1]
            time_since_peak.loc[idx] = (idx - most_recent_peak).total_seconds() /
↪ 3600
        else:
            time_since_peak.loc[idx] = 24 * 7
    df[f'{col}_hours_since_peak'] = time_since_peak
```

Feature Selection and Model Preparation

[34]:
```
# Drop rows with any NaNs created during lag/rolling feature generation
df_clean = df.dropna()

# Define target and exclude original raw columns and categorical ones
target = 'Zone 1 Power Consumption'
exclude_cols = power_cols + ['hour', 'weekday', 'month', 'day', 'quarter',␣
↪'time_of_day']
features = [col for col in df_clean.columns if col not in exclude_cols and␣
↪df_clean[col].dtype != 'object']

X_og = X = df_clean[features]
y_og = y = df_clean[target]

# Fit a Random Forest for feature importance
rf_importance = RandomForestRegressor(n_estimators=40, random_state=42)
rf_importance.fit(X, y)

# Store and display importance
importances = rf_importance.feature_importances_
importance_df = pd.DataFrame({
    'Feature': features,
    'Importance': importances
}).sort_values('Importance', ascending=False)

# Plot top 20 most important features
plt.figure(figsize=(12, 10))
top_features = importance_df.head(20)
sns.barplot(x='Importance', y='Feature', data=top_features)
plt.title('Top 20 Most Important Features for Predicting Zone 1 Power␣
↪Consumption', fontsize=14)
plt.tight_layout()
plt.show()

print("Top 5 most important features:")
print(importance_df.head(5))
```

```
# Select top 20 features for modeling
top_features_list = importance_df.head(20)['Feature'].tolist()
print(f"\nSelected {len(top_features_list)} features for modeling.")
```

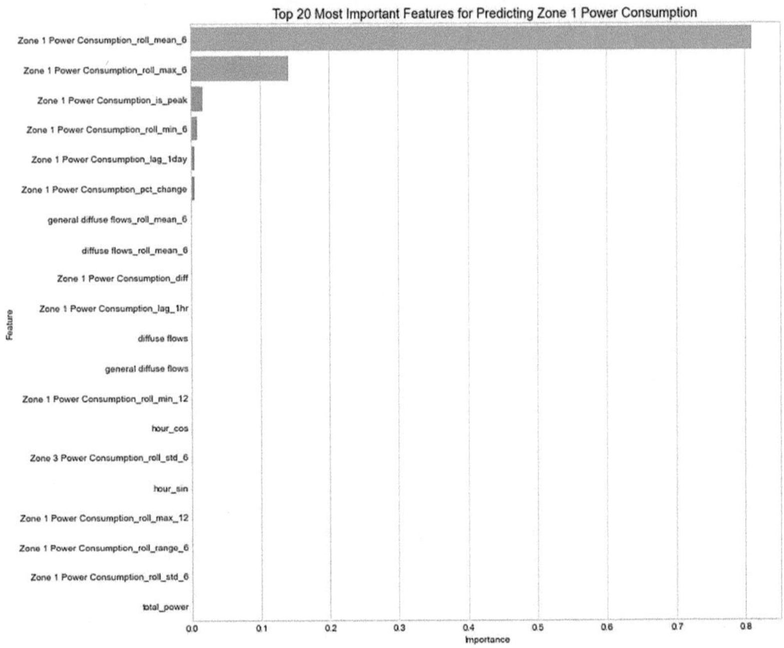


```
Top 5 most important features:
                                Feature  Importance
64   Zone 1 Power Consumption_roll_mean_6    0.810322
67    Zone 1 Power Consumption_roll_max_6    0.140403
164      Zone 1 Power Consumption_is_peak    0.017336
66    Zone 1 Power Consumption_roll_min_6    0.009967
48     Zone 1 Power Consumption_lag_1day    0.005785

Selected 20 features for modeling.
```

Model Training and Evaluation

[35]:
```
# Helper function to evaluate models
def evaluate_model(name, model, X_train, y_train, X_test, y_test):
    model.fit(X_train, y_train)
    y_pred_train = model.predict(X_train)
    y_pred_test = model.predict(X_test)

    metrics = {
        'Model': name,
        'Train R2': r2_score(y_train, y_pred_train),
        'Test R2': r2_score(y_test, y_pred_test),
        'Train RMSE': np.sqrt(mean_squared_error(y_train, y_pred_train)),
        'Test RMSE': np.sqrt(mean_squared_error(y_test, y_pred_test)),
        'Train MAE': mean_absolute_error(y_train, y_pred_train),
        'Test MAE': mean_absolute_error(y_test, y_pred_test),
        'Test MAPE': np.mean(np.abs((y_test - y_pred_test) / (y_test + 1e-5))) *␣
↪100
    }
    return metrics,y_pred_test

# Prepare data
X = df_clean[top_features_list]
y = df_clean[target]
split_index = int(len(df_clean) * 0.8)
X_train, X_test = X.iloc[:split_index], X.iloc[split_index:]
y_train, y_test = y.iloc[:split_index], y.iloc[split_index:]
```

- Let's call our new Linear Regression model containing advanced feature engineering as Advanced Linear Regression.
- We compare our results against all the models from the basic notebook.

[36]:
```
results_df = pd.read_csv('basic_nb_eval_results.csv')

name = 'Advanced Linear Regression'
metrics, y_pred = evaluate_model(name, LinearRegression(), X_train, y_train,␣
↪X_test, y_test)

# Append the metrics dict as a new row to results_df
results_df = pd.concat([results_df, pd.DataFrame([metrics])])
results_df.set_index("Model", inplace=True)
results_df.sort_values("Test R2", ascending=False, inplace=True)

# Show results
display(results_df)
```

```
                            Train R2   Test R2   Train RMSE    Test RMSE  \
Model
Advanced Linear Regression  0.997980  0.998060   319.140468   269.709511
Gradient Boosting           0.362081  0.048759  5729.333537  6014.169853
Lasso Regression            0.192089  0.044084  6447.676267  6028.931950
Ridge Regression            0.192089  0.044084  6447.676267  6028.932342
Linear Regression           0.192089  0.044084  6447.676267  6028.932599
Random Forest               0.947564 -0.223839  1642.615408  6821.700465

                              Train MAE     Test MAE  Test MAPE
Model
Advanced Linear Regression   210.150322   182.464885   0.621432
Gradient Boosting           4507.037797  4774.936183  17.942949
Lasso Regression            5314.320299  5082.744315  18.928821
Ridge Regression            5314.320325  5082.744596  18.928823
Linear Regression           5314.320342  5082.744456  18.928821
Random Forest               1089.681304  5008.637028  18.873822
```

[37]:

```
# Compute % improvements
adv = results_df.loc['Advanced Linear Regression']
gb = results_df.loc['Gradient Boosting']
higher_is_better = ['Train R2', 'Test R2']

improvements = {
    col: ((adv[col] - gb[col]) / abs(gb[col]) * 100) if col in higher_is_better
    else ((gb[col] - adv[col]) / abs(gb[col]) * 100)
    for col in results_df.columns
}

# Print Test R2 improvement
test_r2_improvement = improvements['Test R2']
print(f"Advanced Linear Regression improved Test R² by {test_r2_improvement:.
 ↪2f}% over Gradient Boosting!!!")
```

```
Advanced Linear Regression improved Test R² by 1946.91% over Gradient
Boosting!!!
```

[38]:

```
# Compare other metrics
other_metrics = {k: v for k, v in improvements.items() if k != 'Test R2'}
plt.figure(figsize=(10, 5))
sns.barplot(x=list(other_metrics.keys()), y=list(other_metrics.values()),␣
 ↪palette='viridis')
plt.axhline(0, color='gray', linestyle='--')
plt.ylabel('Improvement (%)')
plt.title('Improvement of Advanced Linear Regression over Gradient Boosting')
plt.xticks(rotation=45)
plt.tight_layout()
plt.show()
```

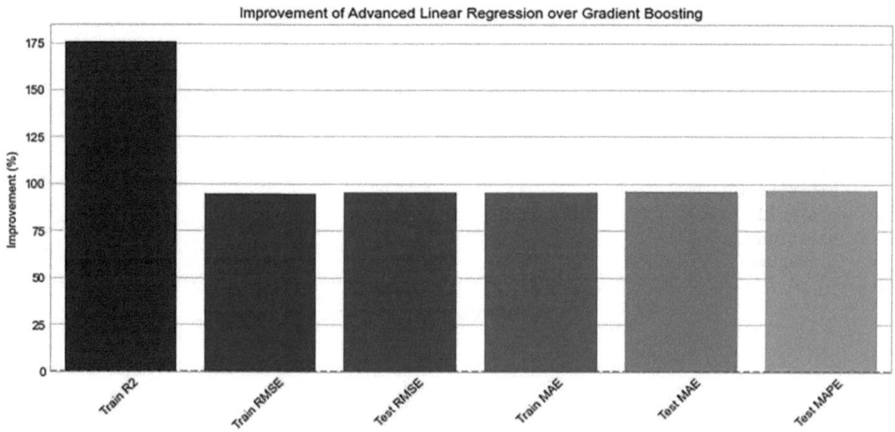

Visualizing Predictions

[39]:
```
# Plot actual vs predicted values over time
plt.figure(figsize=(16, 6))
plt.plot(y_test.index, y_test.values, label='Actual', color='blue', linewidth=2)
plt.plot(y_test.index, y_pred, label='Predicted', color='green', alpha=0.7)
plt.title("Actual vs Predicted Power Consumption (Zone 1)", fontsize=14)
plt.xlabel("Time")
plt.ylabel("Power Consumption (kW)")
plt.legend()
plt.tight_layout()
plt.grid(True)
plt.show()
```

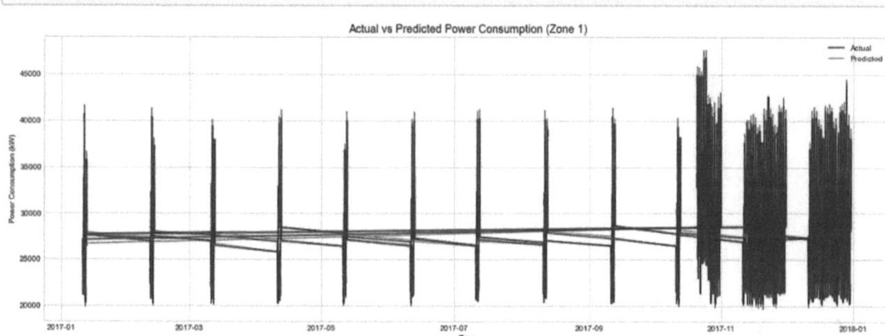

[40]:
```
# Scatter plot of actual vs predicted
plt.figure(figsize=(6, 6))
sns.scatterplot(x=y_test, y=y_pred, alpha=0.5)
plt.plot([y_test.min(), y_test.max()], [y_test.min(), y_test.max()], '--',␣
    ↪color='red')
plt.xlabel("Actual")
plt.ylabel("Predicted")
plt.title("Actual vs Predicted Scatter Plot")
plt.grid(True)
plt.tight_layout()
plt.show()
```

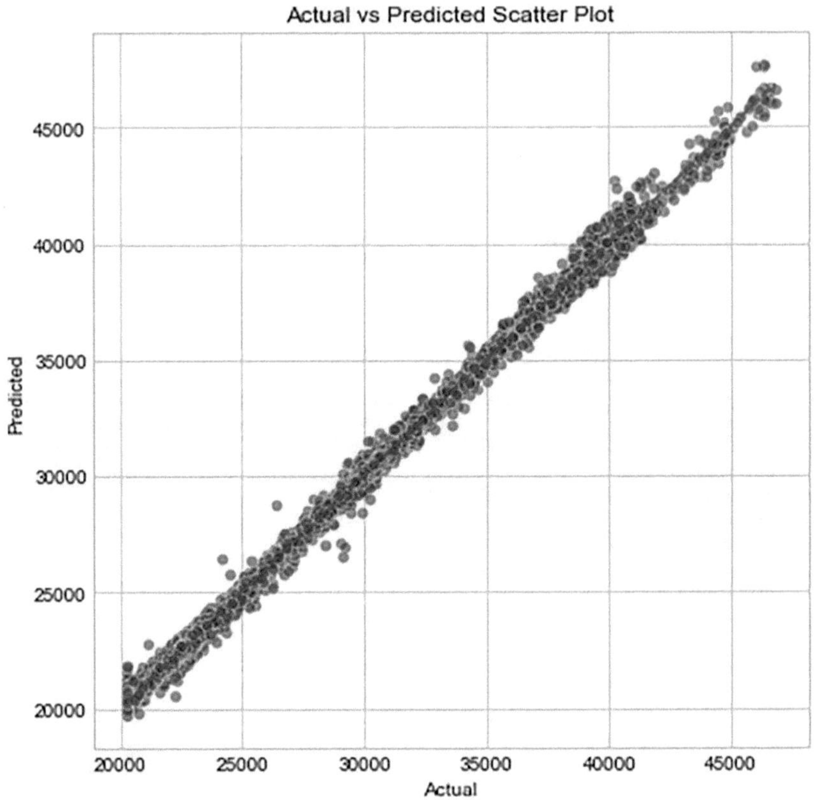

Conclusion

In this notebook, we developed a robust machine learning pipeline to forecast zone-wise electricity consumption in Tetouan City using weather and temporal data. Key steps included:

- **Data cleaning and preprocessing** of the UCI dataset.
- **Exploratory data analysis** covering missing values, temporal patterns, and weather impact.
- **Advanced feature engineering** with lag, rolling, and derived features.
- **Evaluation and visualization** of prediction accuracy.

I encourage you to experiment further using the tools and techniques demonstrated throughout these notebooks—there's plenty of opportunity to explore and improve predictive performance on your own! To get started, here are a few ideas:

- Extend modeling to **all** zones using multi-output regression.
- Incorporate external events and holidays to capture demand spikes.
- Experiment with non-linear models, ensembling strategies, or temporal deep learning (e.g., LSTMs or TCNs).
- Try model interpretability techniques like **SHAP** to better understand driver variables.

Reference

1. A. Salam, A. El Hibaoui, Power consumption of Tetouan city. UCI Machine Learning Repository (2018). https://doi.org/10.24432/C5B034

Chapter 9
Project Management and Implementation

9.1 Key Phases of Renewable Energy Projects

The successful execution of renewable energy projects involves a comprehensive and organized process that spans multiple phases. These phases ensure that projects are carefully evaluated, designed, and implemented efficiently while addressing technical, environmental, and financial considerations. Below, we explore the key phases of renewable energy project development, including feasibility studies, design and engineering, and construction through to operations and maintenance (O&M).

9.1.1 Feasibility Studies and Site Selection

The first critical phase in any renewable energy project is the **Feasibility Study and Site Selection**, which lays the groundwork for all subsequent project activities. This phase involves identifying potential locations for energy generation, conducting technical and environmental assessments, and ensuring that the chosen site aligns with the project's goals [1].

Evaluating Potential Sites During the feasibility phase, renewable energy developers perform thorough assessments of potential sites. These assessments consider factors such as the availability of natural resources (wind, solar, geothermal, etc.), proximity to infrastructure (power lines, roads, etc.), and regulatory conditions. For instance, in wind energy projects, factors like wind speed, consistency, and terrain are analyzed using wind resource assessments, which help determine the site's energy production potential (Baker, 2020). Similarly, solar energy sites require an evaluation of solar radiation levels, cloud coverage, and orientation [2].

D. Gupta et al., *The Roadmap to Renewable Energy*,
https://doi.org/10.1007/978-3-032-09609-8_9

Initial Project Assessments Feasibility studies also involve comprehensive assessments of the site's environmental and social impacts. These assessments typically include evaluations of wildlife, land use, water resources, and potential noise impacts. Additionally, project developers conduct **environmental impact assessments (EIAs)**, ensuring that the project complies with local and national environmental regulations (Sioshansi, 2019). Social impact studies help assess how the project will affect local communities, including factors such as job creation and land use [3].

Finally, financial analysis is conducted to estimate the project's costs and returns, ensuring the project is economically viable and attracts investors. The feasibility study culminates in a decision-making process, determining whether the project should proceed to the next phase [1].

9.1.2 Design and Engineering

Once the feasibility study is complete and the project is deemed viable, the next phase involves **Design and Engineering**. In this phase, the project transitions from a conceptual plan to a detailed blueprint, involving extensive technical design and engineering considerations [4].

Technical Design The design phase focuses on creating a detailed technical plan for the construction and operation of the renewable energy system. This includes selecting appropriate technologies, designing the layout of the energy generation infrastructure, and optimizing system efficiency. For solar projects, this could involve designing the panel array configuration, electrical wiring systems, and energy storage integration. For wind projects, it may involve designing turbine layouts, electrical collection systems, and substations [5].

Engineering Considerations Engineering considerations include structural analysis, electrical design, and the integration of renewable energy systems with existing grids or storage systems. Engineers ensure that the project is built to meet safety standards, optimize energy output, and comply with grid codes (Lund & Mikkola, 2016). Moreover, engineers assess potential risks such as seismic activity, extreme weather events, or logistical challenges to minimize delays and costly redesigns [6].

Project Planning The planning aspect in this phase involves the creation of a detailed project timeline, allocation of resources, and the identification of potential risks and mitigation strategies. Project managers use tools like Gantt charts, risk matrices, and project management software to establish milestones and track progress throughout the project lifecycle [4].

9.1.3 Construction, Operations, and Maintenance (O&M)

The final phase in the lifecycle of a renewable energy project involves **Construction, Operations, and Maintenance (O&M)**, which ensures the transition from the planning and design phase to the operational phase of the project.

Construction During the construction phase, all physical work is performed to build the infrastructure, including the installation of renewable energy equipment (wind turbines, solar panels, etc.), the construction of supporting infrastructure (roads, substations, transmission lines), and the setup of control and monitoring systems. Construction managers oversee the work, ensuring that it progresses according to the project's schedule, within budget, and in compliance with safety standards [5].

Operations Upon completion of construction, the project transitions into the **Operations** phase, where the energy generation system begins to produce and supply electricity to the grid. During this phase, project managers closely monitor system performance, ensuring that energy production aligns with projections. Key operational metrics, such as efficiency, output, and system reliability, are continuously tracked to ensure optimal performance. Operators perform routine checks, monitor environmental conditions, and respond to any equipment failures or performance issues [6].

Maintenance **Operations and Maintenance (O&M)** is a critical phase for ensuring the long-term performance of renewable energy systems. Preventative maintenance is scheduled regularly to minimize system downtime and prolong the lifespan of equipment. This includes inspections, cleaning, lubrication, and parts replacement as needed. Scheduled O&M activities vary based on the type of renewable energy system, wind turbine blades may need inspection and repair, while solar panels may require cleaning to maintain optimal energy generation [2].

In addition to routine maintenance, remote monitoring technologies are often used to detect issues early, reducing response time and avoiding significant disruptions. Maintenance plans are typically structured to balance between scheduled downtime for inspections and maximizing energy production, ensuring that the project remains profitable and sustainable [4].

Transitioning to Long-Term Operations As the project moves into long-term operations, the O&M team may transition responsibility to a dedicated operations team if the project is sold to a new owner or operator. The O&M phase continues for the life of the project, which can span decades. The transition from construction to operations is critical to maintaining performance, and a well-executed plan ensures a smooth handover of control and minimal disruptions to energy production [1].

9.1.4 Stakeholder Engagement and Risk Management

Effective project management in renewable energy development is not only about technical design and construction but also about managing relationships with key stakeholders and identifying and mitigating risks. Stakeholder engagement and risk management are crucial components that directly affect the project's success, from initiation to operations. This section focuses on the importance of engaging stakeholders early in the process, identifying and mitigating risks, and addressing conflicts through clear communication strategies.

Identifying and Engaging Stakeholders One of the most important aspects of managing renewable energy projects is **Identifying and Engaging Stakeholders**. Stakeholders can include local communities, government agencies, investors, environmental groups, regulatory bodies, contractors, and even the end-users of the energy produced. The success of a renewable energy project is often contingent on the involvement and collaboration of these diverse stakeholders from the earliest stages [7].

Importance of Early Stakeholder Involvement Engaging stakeholders early in the project helps establish trust, gather diverse perspectives, and ensure that the project's goals align with broader community interests. For instance, early engagement with local communities can identify concerns such as environmental impacts, land use, and job creation, which are critical for gaining social acceptance [8]. Involving local communities early also increases the likelihood of receiving necessary permits and support from local governments.

Additionally, involving stakeholders from the beginning allows for the early identification of potential challenges, including regulatory hurdles and potential opposition, which can be addressed proactively. This reduces the likelihood of delays later in the project [7]. By establishing a continuous line of communication with stakeholders, project managers can foster a sense of ownership and accountability, encouraging support throughout the project lifecycle.

Collaboration Across Sectors Collaboration is key, as renewable energy projects often involve diverse and sometimes conflicting interests. Effective engagement strategies should include open meetings, public consultations, and participatory decision-making processes to allow stakeholders to express concerns and offer input [9]. For example, environmental organizations may push for stricter environmental protections, while government agencies may focus on policy alignment and economic outcomes. Collaborating with all involved groups ensures that the project benefits from a balanced and inclusive approach.

9.2 Risk Identification and Mitigation Strategies

All projects, particularly in the renewable energy sector, face a range of risks that can impact their success. Identifying these risks early and developing mitigation strategies is essential for minimizing disruptions and ensuring project completion within budget and on time. Renewable energy projects are particularly prone to risks due to their complex, long-term nature, and sensitivity to external factors (Jasanoff, 2010).

Common Risks in Renewable Projects

- **Technical Risks:** These include risks related to the performance of renewable energy technologies, such as wind turbines and solar panels. Risks might arise from technological obsolescence, equipment failures, or system inefficiencies.
- **Financial Risks:** Securing adequate financing is often one of the biggest challenges. Risks arise from fluctuations in energy prices, changing subsidies or incentives, and cost overruns.
- **Regulatory Risks:** These risks are related to changes in policies, laws, and regulations. For instance, shifts away from government incentives for renewable energy or stricter environmental regulations can impact the feasibility and profitability of a project.
- **Environmental Risks:** These risks include issues like land availability, environmental impact assessments (EIA), and unforeseen natural disasters. For example, a site that initially seems viable for wind power may later be shown to have wildlife conservation concerns [6].
- **Social Risks:** Local opposition, community concerns, and stakeholder protests can present significant barriers. Local communities may resist the development of projects due to concerns about their environment, economy, or quality of life [9].

Mitigation Strategies Mitigating these risks involves developing comprehensive risk management plans at the outset of the project. Project managers can mitigate technical risks through careful vendor selection, quality assurance measures, and the use of proven technologies. Financial risks can be addressed by diversifying funding sources and seeking stable government-backed incentives [10].

Regulatory risks can be mitigated by conducting thorough due diligence on applicable policies and keeping abreast of potential changes. This includes engaging with government regulators and ensuring that the project complies with current and future regulations [6]. Environmental risks require a thorough Environmental Impact Assessment (EIA), and social risks can be addressed through transparent community engagement and benefit-sharing arrangements.

Furthermore, the establishment of risk registers, regular risk reviews, and contingency plans helps ensure that risks are continuously monitored and appropriately addressed throughout the project lifecycle [8].

9.2.1 Conflict Resolution and Communication Plans

Effective **Conflict Resolution and Communication Plans** are essential for maintaining stakeholder relations throughout the project. Disagreements and conflicts are inevitable when managing large, multifaceted projects, and how these issues are handled can significantly impact the project's success.

Methods to Address Stakeholder Conflicts Stakeholder conflicts can arise due to differences in priorities, economic interests, or environmental concerns. Addressing these conflicts requires clear communication and a commitment to finding mutually beneficial solutions. Common strategies for conflict resolution include:

- **Mediation and Negotiation:** In cases where conflicts arise, facilitated negotiations or mediation with third-party experts can help resolve issues in a manner that is acceptable to all parties involved. This method fosters cooperation and preserves relationships [10].
- **Transparent Communication:** Open, transparent, and consistent communication is crucial for managing conflicts. Regular updates about project progress, changes in scope, and potential impacts ensure that all stakeholders are informed and have an opportunity to voice concerns before they escalate [9].
- **Win-Win Solutions:** Where possible, project managers should aim for solutions that offer benefits to all stakeholders. This could include financial compensation for affected communities, environmental mitigation measures, or job creation opportunities for local workers [7].

Communication Plans An effective communication plan ensures that all stakeholders, from community members to regulatory agencies, are informed and engaged at appropriate stages of the project. A successful plan includes regular updates, feedback loops, and clear protocols for addressing concerns. For example, monthly newsletters, public forums, and project websites can serve as platforms for sharing information and receiving feedback [8].

Additionally, communication strategies should be tailored to the needs of different stakeholders. Local community members may require more frequent, face-to-face engagement, while investors and regulatory bodies may prefer formal reports and digital communication channels.

Ensuring Transparency Transparency is a fundamental element in managing stakeholder relations and resolving conflicts. Ensuring that stakeholders have access to accurate and timely information helps build trust and reduces the potential for misunderstandings and disputes. Transparency can be further enhanced through third-party audits and public reporting on the project's social, environmental, and financial outcomes [6].

9.2.2 *Permitting, Licensing, and Regulatory Compliance*

Renewable energy projects are subject to a wide range of regulations, permits, and licensing requirements at the local, state, and federal levels. Navigating these regulatory frameworks is crucial for the successful development and operation of a project. Compliance with these regulations ensures that projects adhere to environmental, safety, and economic standards while also reducing the potential for legal disputes and delays. This section explores the key aspects of permitting, licensing, and regulatory compliance for renewable energy projects.

9.2.2.1 Navigating Local, State, and Federal Regulations

Overview of Compliance Requirements at Various Governmental Levels At every stage of a renewable energy project, from site selection to operation, compliance with local, state, and federal regulations is essential. Different levels of government regulate different aspects of renewable energy projects, and understanding how these regulations interact is crucial for avoiding delays and ensuring smooth development.

- **Local Regulations:** Local governments typically regulate land use, zoning, and construction activities. Local authorities assess whether a renewable energy project complies with local land-use plans and building codes. Local permits, such as conditional use permits or variances, may be required for projects depending on their location. These permits are often the first step in the regulatory process and must be obtained before state or federal agencies become involved [11].
- **State Regulations:** State governments play a significant role in renewable energy project development. Many states have adopted renewable energy portfolio standards (RPS) or mandates, requiring utilities to produce a certain percentage of their energy from renewable sources. Additionally, states regulate environmental impacts through agencies responsible for air and water quality, waste management, and wildlife conservation. State-level agencies often issue permits for water use, air emissions, and stormwater management [12].
- **Federal Regulations:** At the federal level, several agencies regulate renewable energy projects, particularly in areas such as environmental protection, wildlife conservation, and energy production. The U.S. Environmental Protection Agency (EPA) is a key federal agency involved in ensuring projects comply with environmental laws. Additionally, the Federal Energy Regulatory Commission (FERC) oversees electricity transmission and distribution systems and may need to approve certain aspects of renewable energy projects, especially those involving large-scale grid connections [13]. Other agencies, such as the U.S. Fish and Wildlife Service (USFWS), may also be involved, particularly in projects that could impact protected species or habitats.

Understanding the layers of regulation at each level of government is crucial for project managers to ensure timely approvals and avoid costly delays. A comprehensive strategy for navigating these regulatory requirements includes early consultation with regulators, legal advisors, and environmental experts to ensure all relevant permits are identified and obtained.

9.2.2.2 Environmental Permitting and Impact Assessments

Key Environmental Permits and Conducting Environmental Impact Assessments (EIA) Environmental permitting and conducting an Environmental Impact Assessment (EIA) are critical components of regulatory compliance for renewable energy projects. These processes ensure that potential environmental impacts are identified, evaluated, and mitigated.

- **Key Environmental Permits:** Renewable energy projects are subject to several environmental permits depending on their type and location. For example, wind energy projects may require permits related to wildlife impacts, while solar projects may need permits for land disturbance or stormwater runoff. Common environmental permits for renewable energy projects include:
  - **National Environmental Policy Act (NEPA) Review:** Under NEPA, federal agencies must evaluate the environmental effects of major federal actions, including renewable energy projects on public lands.
  - **Water Permits:** Projects involving water use or runoff may require permits from state or federal agencies, such as the U.S. Army Corps of Engineers or local water management districts, to ensure that water quality is maintained.
  - **Air Quality Permits:** Renewable energy projects that produce emissions, even indirectly, may require air quality permits, such as those governed by the Clean Air Act.
  - **Wildlife Protection Permits:** Wind and solar projects can have significant impacts on local wildlife. In such cases, the project may require consultation with agencies like the U.S. Fish and Wildlife Service to obtain necessary permits and implement mitigation measures.
- **Environmental Impact Assessments (EIA):** An EIA is a formal process used to evaluate the potential environmental impacts of a project. An EIA involves collecting baseline data on environmental conditions, analyzing the potential impacts of the project on those conditions, and identifying mitigation measures to reduce adverse effects. The EIA process typically includes:
  - **Scoping:** Identifying the key environmental issues that need to be studied.
  - **Impact Analysis:** Assessing how the project will affect various environmental resources, including air, water, land, wildlife, and vegetation.
  - **Public Involvement:** Engaging stakeholders and local communities to ensure their concerns are addressed in the assessment process.

– **Mitigation Measures:** Developing strategies to mitigate identified environmental impacts, such as wildlife protection measures or water management plans [12].

Conducting a thorough EIA and obtaining the necessary permits is essential for minimizing environmental harm and ensuring that the project complies with legal and regulatory standards. This process also fosters positive relationships with local communities and stakeholders, enhancing the project's social license to operate.

9.2.2.3 Compliance with Renewable Energy Standards

Adherence to Industry and Regulatory Standards for Renewable Energy Compliance with renewable energy standards is essential for ensuring that projects meet both industry best practices and regulatory requirements. Various standards and certifications apply to different types of renewable energy projects, such as solar, wind, and storage systems.

- **Renewable Energy Certificates (RECs) and Guarantees of Origin (GOs):** In many regions, renewable energy producers can earn RECs or GOs for the electricity they generate. These certificates verify that the electricity was produced from renewable sources, providing a market mechanism for tracking renewable energy generation. Compliance with these standards allows renewable energy producers to sell their RECs, providing additional revenue streams and contributing to the achievement of renewable energy targets [13].
- **Industry Standards:** Renewable energy projects must adhere to various industry standards that address safety, reliability, and performance. For example, the International Electrotechnical Commission (IEC) provides standards for the design, testing, and performance of renewable energy systems, including wind turbines and solar panels. Compliance with these standards helps ensure that systems are safe, reliable, and efficient [11].
- **Grid Connection Standards:** Renewable energy projects must also comply with grid connection standards set by national or regional grid operators. These standards ensure that renewable energy systems can safely and efficiently integrate with the grid, including requirements for voltage control, frequency regulation, and fault detection [12].

Compliance with these industry and regulatory standards ensures that renewable energy projects meet the technical and operational requirements necessary for success. Adherence to standards not only ensures the project's long-term viability but also helps build trust with stakeholders, including regulators, investors, and the public.

9.3 Project Scheduling and Budgeting

Project scheduling and budgeting are foundational elements of project management in the renewable energy sector. These processes ensure that projects are completed on time, within budget, and to the required quality standards. Effective scheduling and budgeting are crucial for minimizing delays and financial risks, which can otherwise undermine the success of renewable energy projects. This section discusses key strategies for developing project schedules, managing finances, and tracking costs throughout the project lifecycle.

9.3.1 Developing Project Schedules and Timelines

Tools and Techniques for Effective Project Scheduling A well-developed project schedule is critical for ensuring that all tasks are completed on time and in alignment with the project's overall objectives. In renewable energy projects, scheduling includes activities related to site preparation, permitting, design and engineering, construction, commissioning, and operations.

- **Work Breakdown Structure (WBS):** One of the key tools for developing a project schedule is the Work Breakdown Structure (WBS), which breaks down the entire project into smaller, manageable components. This helps to clearly define tasks, deliverables, and milestones. By organizing the work into a hierarchical structure, project managers can ensure that all activities are identified and assigned to appropriate team members or contractors [14].
- **Gantt Charts:** Gantt charts are commonly used to visually represent the project schedule. These charts provide a timeline of all project tasks, showing their start and end dates, dependencies, and milestones. Gantt charts help project managers to track progress and identify any scheduling conflicts or delays early on. They can be created using project management software such as Microsoft Project or Primavera P6 [15].
- **Critical Path Method (CPM):** The Critical Path Method is a scheduling technique used to determine the longest sequence of dependent tasks that must be completed on time for the project to meet its deadline. Identifying the critical path helps to prioritize activities, allocate resources efficiently, and make informed decisions about task dependencies and potential delays [16].
- **Project Management Software:** Utilizing project management software, such as Microsoft Project, Primavera, or online platforms like Asana or Trello, can enhance scheduling by offering tools to allocate tasks, manage timelines, and track progress. These tools facilitate collaboration among team members, helping to keep the project on track and ensuring that everyone is informed about updates and deadlines.

Developing a clear and realistic schedule, using the tools mentioned, ensures that the project adheres to deadlines and can help avoid potential delays that could lead to cost overruns or extended timelines.

9.3.2 Budgeting and Financial Planning

Allocating Resources and Managing Financial Aspects of the Project Budgeting is a critical aspect of managing renewable energy projects, as it ensures that resources are allocated efficiently and the project stays within financial constraints. A well-planned budget helps to avoid overspending, allocates funds for contingencies, and ensures that funds are available for all project activities.

- **Cost Estimation:** The first step in budgeting is accurately estimating the costs for all aspects of the project, including design, engineering, construction, procurement, labor, and operational expenses. Cost estimation techniques such as expert judgment, historical data, parametric modeling, or bottom-up estimating are commonly used to determine the expected costs for each project phase (Pinto [15]).
- **Resource Allocation:** Proper resource allocation is essential for keeping the project within budget. Resources include not only materials and equipment but also human resources, time, and capital. Each resource must be budgeted meticulously to meet the project's needs while minimizing inefficiencies and resource wastage [14].
- **Contingency Planning:** Renewable energy projects often involve a degree of uncertainty, especially due to regulatory approvals, environmental considerations, and weather factors. A good budget includes a contingency fund that accounts for unforeseen circumstances and allows for flexibility in handling unexpected issues. This is especially important for large-scale renewable projects, where delays or cost fluctuations are common [16].
- **Financing the Project:** Renewable energy projects often require significant upfront capital investment, which may be obtained through equity funding, loans, or project financing. In some cases, government grants or incentives may also be available to support the development of renewable energy projects. A sound financial plan that considers various financing options ensures that the project can proceed without delays due to a lack of capital [15].

9.3.3 Monitoring Costs and Avoiding Budget Overruns

Techniques for Tracking Costs and Managing Budget Deviations Monitoring and controlling costs throughout the project lifecycle is essential for ensuring that the project stays within the approved budget. Effective cost management allows project managers to detect potential budget overruns early and implement corrective measures before they become major issues.

- **Earned Value Management (EVM):** Earned Value Management is a technique used to measure project performance and track costs. It compares the planned progress with the actual progress of the project and provides indicators such as Cost Performance Index (CPI) and Schedule Performance Index (SPI) to assess whether the project is on track financially. EVM helps project managers identify cost overruns and take corrective actions promptly [14].
- **Cost Tracking Tools:** Many project management software tools include cost-tracking features that allow project managers to monitor expenses in real time. These tools help track actual costs against the planned budget and flag any deviations. Regular cost tracking provides an early warning system for potential budget overruns, allowing managers to take preventive actions [16].
- **Variance Analysis:** Variance analysis is another tool for monitoring costs. By comparing the estimated costs with actual costs, project managers can identify whether the project is over or under budget. If a variance is identified, the project manager can investigate the causes and take corrective actions. Variance analysis also helps to ensure that the project's financial performance aligns with expectations [15].
- **Change Control Process:** Changes to the scope of work, project timeline, or design specifications can lead to budget increases. A robust change control process helps to evaluate and manage any changes to the project, ensuring that they are properly documented and that their financial impact is understood and accounted for in the revised budget [14].

By consistently tracking costs, managing risks, and addressing issues early, project managers can avoid significant budget overruns and ensure the successful financial completion of renewable energy projects.

9.4 Monitoring and Quality Control During Implementation

Monitoring and quality control during the implementation phase are crucial to ensure that renewable energy projects meet their specified objectives in terms of quality, performance, and safety. These processes help to detect any issues early on and ensure that the project adheres to its planned schedule, budget, and quality standards. This section delves into the strategies for effective construction monitoring, quality assurance, and post-construction testing in renewable energy projects.

9.4.1 Construction Monitoring and Reporting

On-Site Monitoring and Real-Time Project Reporting Tools Construction monitoring involves overseeing the day-to-day activities on the project site, ensuring that all work is proceeding according to the schedule, budget, and quality standards. It helps to detect potential problems before they escalate, providing a timely intervention mechanism to mitigate risks.

- **On-Site Supervision:** Continuous on-site supervision is key for ensuring that construction activities follow the project design and plans. Regular site visits by project managers, engineers, and safety officers help to verify compliance with safety standards, construction specifications, and schedules. This supervision also ensures that the workforce adheres to quality guidelines and best practices [14].
- **Real-Time Project Reporting Tools:** To enhance communication and decision-making, project managers utilize real-time reporting tools such as project management software, mobile applications, and digital dashboards. Tools like Procore, Buildertrend, and PlanGrid allow for real-time tracking of construction progress, change orders, material deliveries, and safety compliance. These tools enable project teams to track project milestones, address delays or issues, and ensure that the project is staying within budget and schedule [16].
- **Construction Progress Monitoring:** Techniques such as **Earned Value Management (EVM)** and **Key Performance Indicators (KPIs)** are used to assess construction progress. By comparing planned progress to actual progress, project managers can identify potential delays, allocate resources efficiently, and adjust schedules as necessary. Regular reporting of these metrics ensures transparency and helps in managing stakeholder expectations [15].

9.4.2 Quality Assurance and Control Procedures

Ensuring Adherence to Quality Standards and Guidelines Quality assurance and control procedures are essential for ensuring that the construction process, materials, and finished systems meet the required standards for performance, safety, and durability. In renewable energy projects, this includes adherence to environmental, engineering, and regulatory standards.

- **Quality Assurance (QA):** Quality assurance involves setting up systems and processes to ensure that the project's design, materials, and construction practices meet the pre-established quality criteria. QA typically involves the creation of a **Quality Management Plan** (QMP) that outlines the procedures for testing, inspection, and documentation throughout the project lifecycle. It also includes selecting qualified contractors and suppliers to ensure high-quality deliverables [16].

- **Quality Control (QC):** Quality control focuses on inspecting, testing, and verifying that construction outputs meet the required quality standards. QC includes regular inspections of construction activities, materials, and installations. For instance, wind turbine foundations, solar panel installations, and electrical systems are all subject to stringent QC checks before proceeding to the next phase. Any deficiencies or non-conformities are identified and corrected before project completion [14].
- **Third-Party Audits and Inspections:** In many cases, third-party auditors or independent quality inspectors are employed to ensure unbiased evaluations of the construction and engineering processes. These audits verify compliance with industry standards such as ISO, ASTM, and specific renewable energy-related certifications. For example, wind projects must comply with IEC 61400 standards, while solar projects must adhere to IEC 61215 and IEC 61730 standards for module safety and performance [15].
- **Non-Destructive Testing (NDT):** Non-destructive testing methods, such as ultrasonic testing, X-ray inspection, and visual inspection, are often used during construction and assembly to detect potential issues in materials or components without damaging them. These methods ensure the longevity and safety of the systems being installed, such as wind turbine blades, solar panel arrays, and structural supports [16].

9.4.3 Post-construction Testing and Commissioning

Testing Systems and Final Project Acceptance Before Full Operations Once construction is complete, it is essential to conduct comprehensive testing and commissioning activities to verify that all systems function as intended and meet operational performance standards. This process also includes formal project acceptance by stakeholders.

- **System Testing:** System testing involves checking all systems, equipment, and infrastructure to ensure they operate as designed. For solar projects, this includes testing the photovoltaic (PV) systems, inverters, electrical connections, and grid integration. For wind projects, it includes testing turbine blades, rotor systems, and the electrical grid interface. These tests are usually performed under controlled conditions to simulate actual operating environments [14].
- **Performance Testing and Monitoring:** During the commissioning phase, renewable energy systems undergo performance testing to assess their efficiency and effectiveness. For example, a wind turbine must achieve its rated capacity under specific wind conditions, while a solar array should perform optimally in varying sunlight conditions. Performance testing involves measuring electrical output, load testing, and system response to various operational scenarios. Data from these tests is compared to performance guarantees and specifications provided by the manufacturer [16].

- **Commissioning Documentation and Final Acceptance:** Once all systems have been tested and verified, the project moves to final acceptance. Commissioning documentation, including test reports, system performance data, and compliance certificates, is compiled and reviewed. This documentation is crucial for demonstrating compliance with regulatory and industry standards. Once all criteria are met, the project is formally accepted, and the renewable energy systems can transition to full operational status [15].
- **Handover to Operations and Maintenance (O&M):** After commissioning, the project is transitioned to the operations and maintenance (O&M) team, who are responsible for managing the plant's performance and availability over its lifecycle. This phase includes specialized training for O&M personnel on technologies such as photovoltaic (PV) systems, wind turbines, and battery storage. Maintenance schedules are finalized for key components like inverters, turbines, and battery management systems, and remote monitoring tools are deployed to track performance metrics such as energy yield, capacity factor, and degradation rates. Ensuring spare part availability and grid compliance are also key aspects of this phase [14].

References

1. T. Ahern, *Renewable Energy Project Development* (Routledge, New York, 2015)
2. H. Lund, J. Mikkola, *Renewable Energy Systems: A Smart Energy Systems Approach to the Choice and Modeling of 100% Renewable Solutions*, 2nd edn. (Academic Press, Cambridge, 2016)
3. J. Glasson, R. Therivel, A. Chadwick, *Introduction to Environmental Impact Assessment*, 4th ed., (Routledge, Abingdon, UK, 2012)
4. N. Baker, *Wind Energy Handbook*, 3rd edn. (Wiley, Hoboken, 2020)
5. P. Gipe, *Wind Energy for the Rest of Us: A Comprehensive Guide to Wind Power and How to Use It* (Chelsea Green Publishing, Vermont, 2014)
6. F.P. Sioshansi, *Energy Storage and the Smart Grid: An Overview* (Academic Press, Cambridge, 2019)
7. R.E. Freeman, *Strategic Management: A Stakeholder Approach* (Pitman Publishing, Boston, 1984)
8. P.D. Burns, D.M. Newbery, W. Vogt, Energy project development: Risk management and stakeholder engagement strategies. Energy Policy **74**, 1–13 (2014)
9. R.K. Mitchell, B.R. Agle, D.J. Wood, Toward a theory of stakeholder identification and salience: defining the principle of who and what really counts. Acad. Manag. Rev. **22**(4), 853–886 (1997)
10. S. Jasanoff, The risks of risk management: challenges to policy and practice in the energy sector. Energy Policy **38**(10), 5440–5452 (2010)
11. S.L. Klein, L.A. Murphy, Navigating local, state, and federal renewable energy regulations. Energy Policy **85**, 45–52 (2015)
12. D. Anderson, M. Brown, R. Henderson, Environmental and regulatory challenges in renewable energy projects. Environ. Manag. **45**(3), 275–289 (2018)
13. P. Smith, S.P. Knight, Regulatory compliance in renewable energy development: navigating the legal landscape. Renew. Energy Law J. **10**(2), 65–79 (2017)

14. H. Kerzner, *Project Management: A Systems Approach to Planning, Scheduling, and Controlling*, 11th edn. (Wiley, Hoboken, 2013)
15. J.K. Pinto, *Project Management: Achieving Competitive Advantage*, 4th edn. (Pearson, London, 2016)
16. K. Heldman, *PMP: Project Management Professional Exam Study Guide*, 9th edn. (Wiley, Hoboken, 2018)

Chapter 10
Future Trends and Opportunities in Renewable Energy

Every environmental problem from plastics in the ocean to greenhouse gas (GHG)-driven climate change is the result of human technology, and technology can be used to solve them all. The current overarching problem is climate change. If this is not addressed, the other environmental problems won't really matter. We are sliding into a period of climate instability from our overheated atmosphere with the Earth now committed to some amount of climate change no matter what, but we must prevent it from becoming even worse. Human civilization can adapt to moderate amounts of climate change, but if it goes to extremes, flooding coastal cities and rendering parts of the planet uninhabitable, it will be a much more difficult problem.

Multiple climate assessments from the United Nations Intergovernmental Panel on Climate Change (IPCC) have declared a "Code Red" emergency for the world on climate. In their 2021 state-of-the-science report, the IPCC noted that the world has warmed by 1.1 °C above pre-industrial levels, and is rapidly heading toward 1.5 °C of warming, a threshold where worsening impacts can be expected. Indeed we appear to have crossed the 1.5 °C threshold in 2024 and we are beginning to see the impacts in drier droughts, more intense storms, widespread wildfires, and killer heat waves. The report also pointed out in unusually strong language that the link between human fossil fuel combustion and climate change is "unequivocal." The goal is to reach a new plateau of climate stability.

To stabilize the climate, the IPCC recommends doing two things: (1) stop burning fossil fuels and making it worse; and (2) reduce the existing CO_2 concentrations in the atmosphere to pre-industrial levels. A lot of very smart and clever people are working on technical solutions for clean energy tech and climate tech, and indeed many such solutions already exist. They have not been widely implemented because of cost, economics, political inertia, resistance to change, and a million and one excuses about why it won't work. Pushback from citizens who are willing to shake up the status quo, forgo business as usual, and hold politicians accountable can overcome this lack of action and move clean energy and climate tech forward.

D. Gupta et al., *The Roadmap to Renewable Energy*,
https://doi.org/10.1007/978-3-032-09609-8_10

10.1 The Next Generation of Renewable Technologies

New renewable energy technologies under development include hydrogen, biofuels, and ocean energy systems. These are discussed in more detail in the individual sections that follow.

10.1.1 Green Hydrogen

Hydrogen is a key to the clean energy future. Hydrogen is the lightest element, and typically occurs as a gaseous, diatomic molecule (H_2). Hydrogen contains no carbon and thus burning it does not produce any CO_2 GHG. It combusts readily with oxygen, creating only water vapor (H_2O) as a combustion product. Although water vapor is a GHG and can trap heat like CO_2, it cycles rapidly through the atmosphere. Contributions of water vapor to the atmosphere from the human combustion of hydrogen are never likely to be anything more than minor given the vast amounts of water in the Earth's natural hydrologic cycle.

Hydrogen cannot be used as a direct replacement for natural gas because it burns with an adiabatic flame temperature approximately 88 °C (190 °F) hotter than methane (2045 °C vs. 1957 °C), requiring the replacement of burner heads on natural gas appliances like furnaces and water heaters to prevent overheating and creating a fire hazard. The flame speed of hydrogen is also nearly ten times faster than methane, making it more explosive. The methane molecule is eight times heavier than a hydrogen molecule, so on a per volume basis it takes about three times as much hydrogen to equal the heating value of a given volume of methane, even though on a mass basis hydrogen has over double the energy value of an equal mass of methane.

Hydrogen molecules are very small, and can permeate into metal along grain boundaries, where they react and form metal hydrides. These compounds weaken bonds, separate the grains, and eventually cause brittle failure of the metal. Carbon steel is especially susceptible to this so-called "hydrogen embrittlement" and steel tanks and pipelines carrying hydrogen risk leakage and ruptures. Despite these concerns, hydrogen shows great promise as a future clean energy resource.

Although hydrogen is the most abundant element in the universe, it is rare on Earth as a stand-alone molecule. The gas is so chemically reactive that it generally occurs only in combination with other elements like oxygen, carbon, or nitrogen. Geologists were surprised recently to discover unbonded or free diatomic hydrogen (H_2) deposits underground in certain rocks. Its existence was not widely known because people were simply not looking for it. Molecular hydrogen appears to be created when groundwater reacts with certain metallic minerals that strip oxygen from the water molecules, leaving free hydrogen to migrate and accumulate like natural gas. The amount of this free hydrogen in the subsurface is unknown, but could potentially be quite large.

To be available as a carbon-free fuel, H_2 is generally manufactured by electrolytic or chemical processes from hydrogen-containing compounds. There are several ways to go about this, but the most common method at present for industrial production uses a process called steam reforming that creates H_2 from natural gas. High-temperature steam (700–1000 °C) under high pressure in the presence of a catalyst reacts with the methane that makes up the bulk of natural gas to produce hydrogen H_2 and carbon monoxide (CO). The CO and steam are then reacted further using a catalyst in what is called the "water-gas shift reaction" to produce CO_2 and more H_2. If the CO_2 is captured and sequestered, the product is said to be "blue" hydrogen, but if it is vented to the atmosphere the process makes "gray" hydrogen. However, in all cases hydrogen is a clear, colorless gas, and the color assignments merely refer to how it was produced.

The most environmentally-friendly hydrogen production method is "green" hydrogen, made by the electrolytic dissociation of water using renewable electricity with zero emissions. The discovery that passing an electrical current through water will split it into its component parts is credited to English scientists William Nicholson and Anthony Carlisle, who used a battery to successfully extract hydrogen and oxygen from water in 1800. A Belgian electrical engineer named Zénobe Gramme invented a machine in 1869 that produced industrial quantities of hydrogen through the electrolysis of pure water with small quantities of acid added to improve conductivity. Hydrogen gas accumulates at the negatively charged cathode, while oxygen accumulates at the positively charged anode. Gramme's hydrogen machine never found practical uses, and he is probably better known for inventing the electric dynamo, which became important for the generation of electricity.

Hydrogen can be used to power vehicles in two different ways. It will combust with oxygen in air, so like gasoline, ethanol, or methane it can be used in internal combustion engines to provide power. However, an internal combustion engine running on hydrogen must be specially designed and constructed for this purpose because of the higher temperatures, faster flame speed, and the potential to embrittle steel.

The largest existing vehicle combusting hydrogen is the Space Transportation System, or STS, a large rocket designed by NASA for a return to the moon. The hydrogen fuel onboard this rocket is carried as a cryogenic liquid that must be maintained at the extremely cold temperature of −253 °C or −423 °F. This is obviously impractical and dangerous for automobiles.

A second and perhaps far more useful application of hydrogen for transportation is to use a device known as a "fuel cell" to provide portable, direct electric power in place of a battery in an electric vehicle. Fuel cell technology uses a relatively slow chemical reaction between hydrogen and oxygen (in air) to produce electricity. The U.S. Department of Energy has been supporting research on fuel cell-powered electric vehicles (FCEVs) for many years. Fuel cells consist of negative and positive electrodes embedded in an electrolyte. These vary depending on the operating temperature of the fuel cell and the chemistry of the input gases, but the most common type is the solid oxide fuel cell. These electrolytes tend to be both stable and durable, and solid oxide fuel cells have in fact been used successfully since the 1960s to supply electric power on a variety of spacecraft.

Hydrogen is fed to the negative cathode, and oxygen (or air) is fed to the positive anode. The hydrogen molecules are ionized into protons and electrons by a catalyst at the cathode. The protons migrate through the electrolyte to the anode, while the electrons flow through an external circuit, creating an electric current. The protons and electrons unite with oxygen at the positive anode to produce water and heat. The only combustion product from a fuel cell is pure water, which was actually used for drinking water on various manned spacecraft. Hydrogen FCEVs have a range similar to gasoline-powered vehicles, which could provide a clean energy solution to drivers with so-called "range anxiety" who panic if they don't have a driving range of 400 miles even if they are only going 40 miles on a trip. Fuel cells also operate efficiently at temperatures well below freezing, an advantage over battery-powered electric vehicles, which often perform poorly in the cold.

FCEVs are under development in China. A Hong Kong-based company called Sinosynergy is currently producing four different sizes of commercial fuel cells in Guangdong Province to power buses and trams. They intend to expand to passenger cars, trucks, and ships, and have even produced a fuel cell stack to provide backup power for 5G telecommunication base stations. The company claims to be able to manufacture 20,000 hydrogen fuel-cell stacks per year and is one of the world's largest producers. Clean energy policies, lower costs for hydrogen technology, and strategic planning may grow the global market for hydrogen fuel-cell vehicles to US$20 billion by 2030, according to some assessments.

The Chinese government released the country's first-ever long-term plan for hydrogen energy in 2022. The hydrogen industry is supported by Chinese leadership, who see it as a technology of the future. Sinosynergy receives significant government support and subsidies, and the endorsement of hydrogen by Chinese leaders prompted at least 23 provincial governments to issue similar development plans for hydrogen energy and fuel-cell vehicles.

Hydrogen fuel cell vehicles are available in the U.S., but the only manufacturers that meet current U.S. automobile safety standards are Hyundai and Toyota. The Toyota FCEV is called the Mirai, and it is currently produced as a sedan (https://www.toyota.com/mirai/) although they are also developing prototypes of small pickup trucks. The Toyota website claims the XLE model has a range of about 640 km (400 miles) on a fully-charged hydrogen tank, which is two to three times greater than the driving range of battery-powered electric vehicles on a full charge, and equivalent to that of a gasoline-powered automobile. The onboard tank holds compressed hydrogen under a pressure of 87.5 MPa (12,500 psi), and the manufacturer claims it can be refilled in about five minutes.

According to the USEPA website, there are about 13,000 hydrogen FCEVs being driven in the United States at present, mostly in California where they are used to meet clean air standards. This represent less than 0.005% of the total vehicles on American roads. The U.S. is far behind the Chinese on FCEV development and the hydrogen infrastructure needed for refueling. Drawbacks of fuel cell vehicles include difficulties in obtaining hydrogen and the high cost of the fuel when it can be obtained. This has been a problem in China where hydrogen is some 16 times more expensive than gasoline on a per kilogram basis. Getting hydrogen infrastructure in place is a major challenge for fuel cell-powered vehicles. As mentioned earlier, the transport of hydrogen gas through existing steel pipelines is problematic because of its tendency to embrittle steel and cause fractures. Green hydrogen made by water electrolysis using wind or solar electricity can be manufactured near markets for the fuel, widening distribution and avoiding transport issues.

One option being considered for hydrogen transport is to combine it with nitrogen and create ammonia (NH_3). There are a number of advantages to this. First, ammonia contains a significant amount of hydrogen, about 17.6% by weight. Secondly, ammonia liquifies at the cold but not cryogenic temperature of $-33\,^{\circ}$C ($-28\,^{\circ}$F) under atmospheric pressure and can be transported as a liquid under refrigeration. It can also be dissolved in water for transport. The industrial production of ammonia is an intense, high-energy process that can create significant greenhouse gas if fossil fuel is involved. It is normally produced by the reaction of nitrogen and hydrogen in the presence of iron catalysts at temperatures around 400 to 600 °C and under pressures of 200 to 400 atmospheres. Extracting the hydrogen from the ammonia is the reverse of the synthesis reaction but requires even more heat, with temperatures that may reach up to 1000 °C. Ammonia is both caustic and hazardous in its concentrated liquid form, which is the most efficient method for hydrogen transport. The National Institute for Occupational Safety and Health (NIOSH) in the United States classifies it as an extremely hazardous substance and facilities that produce, store, or use ammonia in significant quantities are subject to strict reporting requirements.

Ammonia was briefly considered by DOE as a potential storage medium for hydrogen onboard FCEVs, but even trace levels of the substance can degrade the electrolyte membranes on fuel cells and "poison" the cell. The idea was abandoned after it was determined that it would be nearly impossible to remove every last trace of ammonia from the stored hydrogen gas in an onboard fuel cell system.

A novel concept for hydrogen transport comes from the Fraunhofer Institute for Manufacturing Technology and Advanced Materials (IFAM) in Dresden, Germany. They have developed a paste of magnesium hydride stabilized with an ester and metallic salt. The paste reacts with water to release hydrogen at a controlled rate that can then be fed into a fuel cell or used for combustion. The paste is stable to temperatures as high as 500 °F (260 °C) and can carry a significant volume of hydrogen. Shell Oil and the Norwegian renewable energy company Norsk Hydro have an agreement to use Norsk's existing renewable energy hubs to generate green hydrogen. The initial goal is to provide green hydrogen as a clean-burning fuel for their own operational needs but they plan to eventually scale-up and produce green hydrogen for the European market. Drilling company Baker Hughes has invested in the Canadian firm Ekona Power to tap into the hydrogen market. Ekona has developed a natural gas reforming process that uses a thermal treatment called pyrolysis to break apart methane molecules inside a reactor vessel, producing hydrogen gas and solid black carbon. The production of solid carbon instead of the CO_2 created during the traditional steam reforming process drastically reduces CO_2 emissions.

Hydrogen is in its early stages, but is likely to draw interest from the fossil fuel industry as FCEVs become more prominent on world markets. Energy is energy after all, and if these companies want to survive the clean energy transition, they will need to be willing to jump onto new energy technologies. Vehicles require onboard energy, and replacing petroleum with something else, be it batteries or hydrogen, is going to require the recovery of minerals and fluids from the deep Earth. Petroleum geologists are already doing this with oil and gas, and it makes sense for them to transition into new Earth resources. Rather than holding onto century-old fossil fuel technology for dear life, oil-dependent regions can drive the transition to sustainable, clean energy renewables like hydrogen. The Kingdom of Saudi Arabia, for example, has access to two saltwater bodies and abundant amounts of sunlight across hundreds of kilometers of desert. One way they could transition away from petroleum is to install solar power arrays in coastal regions along the Red Sea and Persian Gulf to manufacture and bottle green hydrogen from seawater as an energy export. The Saudis could also make blue hydrogen from their abundant natural gas resources and sequester the resulting CO_2 in depleted oilfields. The fossil fuel-based economy is coming to an end, and countries and regions that switch energy production to renewables will more readily adapt into the new reality.

10.1.2 Bioenergy and Advanced Biomass

Biofuels are made from green plants that remove carbon dioxide from the atmosphere via photosynthesis and turn it into biomass. When burned, the CO_2 emission are simply returning the gas back to the atmosphere from whence it came; thus they are considered to be carbon-neutral. If the carbon dioxide is captured and sequestered instead of vented, the biofuels become carbon negative, actively reducing the total CO_2 concentration in the atmosphere. This process is called Biofuel Energy with Carbon Capture and Sequestration or BECCS. The economic incentive for BECCS is that it can generate revenue from the sale of electricity to cover the cost of capturing carbon emissions. It has been promoted as a profitable way to reduce carbon dioxide levels in the atmosphere, although to date it has not been widely adopted. Biofuels can be made from a variety of plants including microbes, algae, grasses, trees, and brush.

The easiest and simplest biofuel to make is methane, which is the main component of natural gas. Biogas methane can be directly substituted for natural gas in utility distribution systems without any modifications because methane is methane. Biological methane can be burned exactly like natural gas to provide energy. The bio-methane is created as a metabolic byproduct by a type of anaerobic bacteria called methanogens. These microbes occur almost everywhere and are found in accumulations of organic matter. In the absence of oxygen, they carry out a digestion-like process that produces the gas. Since the microbes do all the work, the only things humans need to supply for biogas are some large, sealed digestion tanks that can provide the proper anoxic environment, a supply of organic waste material, and a system to collect the gas.

Given the prevalence of combined cycle, natural gas-fired power plants to produce electricity in the U.S., replacing the natural gas supply with biogas could significantly decarbonize the electricity sector while requiring minimal to zero modifications of existing electrical infrastructure. Natural gas is composed of about 99% methane, and biogas is nearly pure methane. Biogas methane can be put directly into the natural gas distribution system as a net zero substitute fuel.

Other biofuels such as ethanol made from corn or wood harvested from trees can also be burned to provide energy, of course. However, the dilemma of harvesting food crops like corn or building materials like wood to make biofuel has constrained development and use. One of the major advantages of methane biogas is that the methanogen bacteria will thrive on virtually any type of organic material, including a lot of waste material that has no other use. Biosolids from wastewater treatment plants, animal manure, agricultural waste, and just about any other source of organic carbon can be turned into methane. The microbes will happily eat it all. The prodigious amounts of manure produced as a byproduct of confined animal feeding operations, a common method for raising livestock in the United States, is an environmental hazard. Many farmers who produce cattle, swine, or poultry are stuck with mountains of waste. This organic material has little use as a fertilizer or soil amendment, contaminates streams and groundwater, and has created anoxic "dead zones" in the Chesapeake Bay and the Gulf of Mexico. If all of this manure was funneled into biogas tanks and converted into usable methane fuel instead, it would substantially reduce the volume of the waste along with providing a clean energy source. Farmers could turn a financial liability into an income-producing asset. Municipal wastewater treatment plants could also benefit from methanogen gasification to help reduce the volume of biosolids waste material that requires ultimate disposal and potentially make some income off energy production.

According to the Energy Information Administration, the United States produced 36.4 trillion cubic feet (TCF) of natural gas in 2022 [1]. The bulk of this (nearly 40%) went for electric power generation, and another significant volume (26%) went to industrial uses. About 7% was exported, while the remainder went to commercial and residential uses. The vast overproduction of shale gas through fracking has produced a glut of cheap natural gas that has made it difficult for biogas to compete economically except in some rare instances. Although currently abundant, natural gas is a finite resource and all natural gas wells will decline in production over time and eventually cease. The electric power industry has suffered in the past when the abundance of natural gas they were counting on for power generation turned into a shortage and they were stuck with megawatts of idle equipment. The shale gas boom will end eventually as well. Forward-looking companies can secure a sustainable supply of methane for their gas-fired power plants by supporting the development and scale-up of biogas. As long as some form of organic waste matter is available, a biogas plant can continue to manufacture methane indefinitely. Genetic engineering of the microbes could improve the rates of methane production. A biogas facility located adjacent to a combined cycle power plant could provide a secure source of fuel that can be made onsite as needed. This will insulate electric utilities from shifts in the cost of natural gas, which can vary widely with changes in supply availability.

Digester tanks for methanogens may need to be scaled up to possibly stadium-sized dimensions to produce a sufficient amount of biogas to run a nominal 500 MW powerplant. However, given the resources that fossil fuel production companies are willing to sink into hydraulically-fractured gas wells in shale formations, deep ocean drilling, and petroleum production in barely accessible locations like the North Slope in Alaska, all of which are quite expensive, the financial risk of building large biofuel facilities near power plants is small.

Liquid biofuels are not well-suited to electrical generation. Fuels like ethyl alcohol (ethanol) are created as the byproduct of a fermentation process, but unlike methanogens, the fermentation microbes require specific sugars and starches to make alcohol. Liquid biofuels are mainly of interest to the transportation sector because their portability is ideal for vehicles that are required to carry onboard fuel. Most commercial gasoline blends now contain at least 10% ethanol, diesel vehicles can run on vegetable oils, and sustainable aviation fuel (SAF) derived from biomass is being increasingly adopted by the airline industry.

SAF can be made from a variety of biomass feedstocks, including municipal solid waste, wood and brush, and vegetable oils and fats. It is manufactured through a complex chemical reaction process similar to that used to produce liquid synfuels from coal, although by using biomass instead of coal it is much closer to carbon-neutral. SAF can be run in existing jet engines when blended with conventional Jet A fuel at ratios of up to 50%. Newer engine designs are in the works that will be capable of running on 100% SAF. Globally, aviation accounts for about 12% of all greenhouse gas emissions from the transportation sector and 3% overall. The international aviation industry has set a goal of 2050 for reaching net zero carbon emissions. SAF will help.

10.1.3 Ocean and Tidal Energy

Emerging technologies in marine energy designed to harness ocean power include tides, ocean thermal energy conversion (OTEC), and the use of submarine turbines. Many of these concepts have been around for a while, but issues with technology, location, and scale-up have limited implementation. However, with the potential emergence of a hydrogen-based economy, ocean energy options along with solar and wind could provide power to manufacture hydrogen from seawater on ocean-based platforms, possibly re-purposed oil drilling rigs. The hydrogen could be stored aboard large tanker ships that would periodically bring it to shore. Tidal power uses the force of the incoming and outgoing tide to rotate a turbine that runs a generator. It requires an impoundment like a conventional hydropower dam, but in this case, the flow runs in both directions: inward during a flood tide, and outward during an ebb tide. Where tides are especially high, such as Canada's Bay of Fundy with a tidal range of up to 15 m (50 ft.), a tidal power plant can efficiently produce substantial amounts of electricity. Even in lower tidal ranges, a bay or sound with a limited or narrow tidal inlet can produce a significant current of moving water that can run a turbine. The Inland Bays of Delaware, for example, have only a single narrow inlet at Indian River. The tidal current flowing through this inlet is quite fast, sometimes even producing whitewater rapids under the Route 1 bridge. Placing a turbine here could generate substantial amounts of electricity. Favorable conditions for tidal energy are generally limited to specific geographic configurations, but where these exist the opportunity to generate power is significant.

The OTEC technology seeks to utilize the temperature difference between warm surface waters and cold bottom waters in the deep ocean to drive a thermocouple-type device to produce electricity. Thermocouples work by connecting two wires of different metals in a circuit, and exposing each junction to a different temperature. The small voltage produced by the temperature gradient is called the Seebeck effect, named for the German physicist who discovered it in 1821. The current produced by an individual thermocouple is typically very small, and the most common use for these devices is to monitor the voltage to measure the temperature of kilns, flames, engines, and other industrial processes. The use of thermocouples to produce power requires that dozens of them to be connected in a series called a thermopile, with half of the connections exposed to heat, and the other half exposed to cold. In an OTEC thermopile, the cold connections would be placed kilometers deep in the ocean where the temperature of the bottom waters is just above freezing, The warm connections would be placed in shallow water, or possibly on land. The Israelis have experimented with asphalt-lined ponds along the shore to collect solar heat for the warm junctions to increase efficiency. Although the OTEC concept has been tested several times, unless the deep ocean waters are close to land, the losses to resistance of running a small current through many kilometers of wire are prohibitive. Even in the best case scenario, the capital expenses are high and the production of power is low. Scale-up to the levels needed for utility grade power has been challenging.

The ocean current turbine option places an underwater turbine that turns a generator in a strong ocean current. The concept has actually undergone some field tests by the Japanese and appears workable. The prototype device is called Kairyu and resembles a submerged airplane, with two propeller-like water turbines and a central unit that houses a buoyancy control system. It will be anchored to the sea floor at 30–50 m of depth (100–160 ft.) in the Kuroshio Current, one of the world's strongest that runs along the eastern coast of Japan. Power will be transmitted to shore via undersea cables. If the Japanese can make this work, it could produce hundreds of gigawatts (GW) of electricity, replacing virtually all of their fossil fuel power plants.

10.2 De-carbonization of Industry

The industrial processes that produce the most substantial greenhouse gas emissions include the manufacturing of concrete, steel, and chemicals. Emissions can be reduced by switching energy sources from fossil fuel to renewable electricity or biomass combustion, and by capturing and storing any carbon dioxide that is produced. In many cases, these alternatives are more expensive than fossil fuel options, and industry has been generally reluctant to adopt them. There is also a certain degree of conservatism in industry that is resistant to trying new ways of doing things. They tend to prefer the status quo, where they know the costs, understand how the processes work, and have long experience with the performance and durability of the products.

The decarbonization of industry will have to address these issues to be successful. Platitudes about environmentally responsible concrete manufacturing or steelmaking to protect the climate will not gain much traction with industry. Protesting outside a steel mill is easy, but the hard work is figuring out how to produce emissions-free steel that performs as well or better and costs the same or less than steel made by existing methods. If it doesn't cost the same and work as well, they won't be interested.

This is the challenge for environmentalists. The economic approach can and has been successful many times. One recent example is the changeover of most fossil fuel electricity production in the United States from coal to natural gas. When natural gas became abundant in the first decade of the twenty-first Century thanks to the successful development of shale gas resources, the electric power industry switched almost wholesale from coal over to gas. This turned out to be good for the climate because natural gas emits only about a third of the CO_2 compared to the equivalent energy obtained from burning coal. Gas is also much cleaner to burn than coal, and does not produce the toxic metals, sulfur compounds, or solid ash from coal-fired power plants.

As a result, greenhouse gas emissions from the United States as a whole peaked at about 6 megatons (MT) per year in 2005, and have since leveled off at about 5 MT annually. Some of this million-ton reduction has resulted from the addition of renewables like wind and solar, but much of it was due to the switchover from coal to natural gas. The coal industry attributed their drop in sales to an imagined "War on Coal" waged by the Obama EPA, but in reality shale gas was to blame.

So, had the electric power industry suddenly become environmentally-responsible and altruistic? Not hardly. The changeover from coal to gas was strictly economic. Combined cycle natural gas power plants produce about 50% more electricity per unit of fossil fuel energy compared to coal-fired power plants. The ability to power a Rankine steam cycle using the hot exhaust from an up-front Brayton cycle gas turbine was such a huge improvement in efficiency that the electric power industry could not pass it up. Interestingly, they have barely mentioned the fact that gas-fired electricity is much better for the climate and the environment than coal. But even if they did the right thing with the wrong motives, it is still the right thing. This is a model for the decarbonization of industry that could work across the board.

10.2.1 Electrification of Industrial Processes

Transitioning from fossil fuels to renewable electricity in industries has been a challenge. Some metallurgical industries, like aluminum smelting, are done most efficiently with electricity. Substituting renewables for fossil energy in this case is simple and straightforward, and no one at the aluminum plant will probably even notice. Steel, on the other hand, is primarily made with fossil fuels to provide heat for the smelting process and carbon as an oxygen scavenger.

The production of steel requires that the iron oxides making up iron ore have the oxygen stripped out in an energy-intensive process called reduction to produce pure iron. Iron binds tightly with oxygen and a material called a reductant is added that binds to the oxygen more tightly than the iron and carries it away. In standard steelmaking, carbon is the reductant of choice in the form of coke, which is made from coal with most of the volatiles removed. The iron ore, coke, limestone and a few other ingredients are heated until molten, usually with fossil fuel, and then placed into a blast furnace. This facility blows pressurized air through the molten iron ore to ignite the coke and encourage it to scavenge oxygen away from the iron. The oxygen is carried off as carbon dioxide or carbon monoxide gas, which are typically vented to the atmosphere. The iron left behind in the blast furnace goes through a subsequent process to make steel. A precise amount of carbon, manganese, and silicon are added to the molten iron to create an alloy metal that cools to become a strong but malleable steel.

One of the more perplexing questions faced by those who want to eliminate fossil fuels from our civilization is "How do you make steel?" Fortunately, there are ways. Direct-reduced iron has the oxygen removed chemically from iron ore in the solid state, without melting. The reducing agents are gases, usually carbon monoxide and hydrogen that scavenge the oxygen away from iron ore pellets, leaving pure iron behind. Both direct-reduced iron and scrap steel can be melted down and converted to new steel in the electric arc furnace. This furnace uses electricity to produce the heat needed to melt iron, and the electric power can be provided by renewables like wind turbines or solar panels. An electric arc furnace can also make new steel from scrap metal. Scrap steel can be readily recycled into new products without having to undergo the oxygen-removal process.

The chemical industry is the largest industrial consumer of oil and gas, about half of which is used for feedstock. It also consumes a lot of energy. The increasing demand for a vast array of chemical products like plastics and fertilizers is not making this easy to solve. The United States, the Middle East, and the Asia Pacific region produce the bulk of the so called "high-value chemicals" that are used as feedstocks for plastic manufacturing. These include ethane, benzene, toluene, and mixed xylenes derived from petroleum. The use of petrochemicals to manufacture plastics does not directly affect the climate as long as they don't end up in the atmosphere as a greenhouse gas. Plastics have plenty of other environmental problems, however, such as single use items filling landfills and great floating mats of plastic trash in the oceans. Electrification of the chemical industry will reduce energy consumption and significantly reduce CO_2 emissions while providing more efficient processes for the manufacturing of chemicals. Such a move will also help the industry to meet more stringent air quality standards.

Most other industrial processes can also be electrified. This is why converting the electrical system to carbon neutral or carbon zero renewables is important for decarbonizing the rest of the economy. Once the electric power sector is decarbonized, everything else can follow.

10.2.2 Carbon Capture, Utilization, and Storage (CCUS)

Carbon capture and storage or CCS is an emissions mitigation technology designed to reduce or prevent CO_2 from a fossil fuel combustion process from entering the atmosphere. It is usually limited to removing carbon dioxide emissions from fixed sources like power plants or industrial sites because much of the apparatus needed is somewhat cumbersome and bulky. CCS has not been widely deployed on fossil fuel power plants to date because of the cost. The U.S. Department of Energy (DOE) goal for CCS on fossil fuel power generation is 95% carbon capture with only a 15% increase in the cost of electricity. The reality is that 90 or 95% carbon capture on a coal-fired power plant essentially doubles the cost of electricity, raising it to levels equivalent to those from a nuclear power plant. Despite the costs, carbon capture has been installed on some coal-fired power plants in Texas, Louisiana, and Wyoming. The carbon dioxide gas is marketed to nearby oilfields for enhanced oil recovery operations, which greatly improves the economics but doesn't do much for the climate.

Because the cost of CCS on coal-fired electricity is so high, DOE came up with the idea of finding other markets for carbon dioxide besides enhanced oil recovery that would improve the economics of carbon capture. The CCS acronym was changed to CCUS, for carbon capture, use, and storage. Both are used in the literature and it is important to note the difference. Not many markets for captured carbon dioxide have been identified in the intervening years, mainly because the few existing industrial users were already supplied by established carbon dioxide production wells. Efforts to use CO_2 as a chemical feedstock to create a so-called "circular economy" have been stymied by competition with the lower cost and more familiar feedstocks derived from petroleum and natural gas.

Some people object to CCUS in general because it was originally developed to enable the electric power industry to continue burning coal. While this is true, it can also be applied to any carbon dioxide emission source, including concrete manufacturing and steel milling. The bottom line is that having CCUS on a fossil fuel power plant is more beneficial for the climate than not having it. Electric power companies and certain industries will continue to burn fossil fuels for decades to come no matter how much the environmental community protests. Some folks may find it to be immoral but so far it's not illegal. In this case, isn't capturing the CO_2 and keeping it out of the atmosphere a better outcome than just letting it waft away?

Although many people think of CO_2 as a chemically-inert gas, it does react with various substances. The most common industrial carbon capture process uses a class of organic chemicals called amines derived from ammonia. The chemistry between CO_2 and amines is an acid–base equilibrium reaction where the amine reacts with CO_2 to form what is called a carbamic salt [2]. The carbamate bond in the salt can be broken by heating or by exposure to low pressures, releasing the CO_2 for storage and recharging the amine for re-use. These processes are known as temperature-swing and pressure-swing adsorption respectively.

Carbon dioxide removal from the air using solid sorbents has been known for decades but was generally applied only in situations where breathing air is limited, such as on submarines and spacecraft. The original method developed during World War II for scrubbing CO_2 out of the air on submarines is called "carbonation." It used chemicals that react with the CO_2 and bind to it chemically. The two most common processes use either calcium hydroxide or sodium hydroxide as the reactants. The spent chemicals can simply be replaced with a fresh batch, or the reaction can be reversed using heat or other means to "recharge" the reactant for re-use. The carbonation reaction with calcium hydroxide is as follows:

$Ca(OH)_2 + CO_2 \rightarrow CaCO_3 + H_2O$

The end products of the reaction are water and calcium carbonate that immobilizes the carbon. $CaCO_3$ is the formula for the mineral calcite, which is the main component of limestone. Limestones are extremely stable and have been around for hundreds of millions of years. A similar carbon capture process uses sodium hydroxide as the reactant:

$NaOH + CO_2 \rightarrow NaHCO_3$

The end product of this reaction is sodium bicarbonate, more commonly known as baking soda. It is more soluble in water than calcite and less suitable for long-term sequestration, but as a process for carbon capture it is quite effective. Once captured from a smokestack or directly from the atmosphere, carbon dioxide must be kept isolated or sequestered from the atmosphere for long time periods to have a positive effect on climate. The minimum isolation period is generally a century, but the longer the better. This time span is known as durability, and it is an important consideration for CCUS.

The term "sequestered" in American English usage means separated or isolated. It is most commonly heard when a jury is said to be sequestered in a criminal trial to keep them isolated from outside news sources and opinions that could potentially influence their decisions. In French the term "sequester" has a different context, where it implies being held against one's will, similar to kidnapping. Given the international nature of the CCS program, DOE did not want to offend French-speaking partners with what was thought to be a rather harsh term for isolating CO_2 and began calling the carbon dioxide isolation process "storage" instead.

CO_2 gas at the standard temperature of 25 °C or 77 °F and a standard pressure of one atmosphere has a density of 1.836 kg/cubic meter. Thus, one metric ton of the gas under standard temperature and pressure conditions occupies a volume of 545 cubic meters (19,235 cubic feet) and can be visualized as a sphere some 10 meters (33 ft.) in diameter or a cube about 8 m (26 ft.) on a side. The volume of CO_2 can be reduced by compressing it as a gas, converting into a supercritical fluid, or storing it as solid dry ice. The amount of space needed for carbon dioxide storage is very dependent on how that storage is done.

There are three main options for keeping captured CO_2 isolated from the atmosphere for long time periods: (1) geologic isolation deep underground in rock pore space, (2) terrestrial sequestration on the Earth's surface as a solid, and (3) ocean sequestration dissolved in deep seawater.

Geologic sequestration: Injecting captured CO_2 deep underground into rock pore space or fractures will keep it isolated from the air. Candidate rocks are depleted oil and gas fields, unmineable coal seams, depleted shale gas wells, deep saltwater aquifers, and volcanic basalts. Each of these has advantages and disadvantages. Successful field experiments have been carried out in all of them except shale gas wells, although the performance of shale for CO_2 storage has been tested in the laboratory [3].

Favored options at present are depleted oil and gas fields and volcanic basalts. Depleted oilfields have existing wells that can often be used for CO_2 injection, saving considerably on the cost of drilling new wells for access. Even if old wells need a workover and new casing or cement, this is still much cheaper than drilling a new well from the surface. Although, maintaining well integrity indefinitely to prevent CO_2 leakage remains a key challenge, increasing costs due to ongoing monitoring and maintenance. Despite this, using existing wells is generally still cheaper than drilling new ones. Oil reservoirs also have porous rock that held the oil, and it will also hold CO_2. The oil and gas were trapped in the underground reservoir by a geologic structure such as a fault or a fold, and contained by an impermeable caprock such as a salt bed. These geologic properties are intrinsic to the particular oil reservoir, and if they held the oil and gas in place for geologic time periods, they will also hold carbon dioxide. Another advantage of depleted oilfields is that they often have reams of information available, including well logs, geophysical surveys, drill core, and production data.

The reservoir rock properties and the shape of the trap and seal are usually well-characterized, and these data can be helpful for planning a CO_2 injection program. Basalt is an extremely abundant volcanic rock formed from solidified lava. It is sourced from the deep crust or upper mantle, and contains a suite of minerals that are unstable on the Earth's surface and break down or "weather" readily when exposed to water and air. When aqueous solutions of CO_2 are injected into basalt, the water and dissolved gas react with the minerals to release metal ions into solution. These cations then combine with carbon dioxide to precipitate carbonate minerals out of solution within the pores and fractures of the basalt [4]. This is an extremely stable method for storing carbon dioxide because once it forms a solid mineral, it is immobilized for geologic time.

Basalt rocks are typically composed of plagioclase feldspar and pyroxene minerals. Plagioclase is a mixture of the calcium end member anorthite, $Ca(Al_2Si_2O_8)$ and the sodium end member albite, $Na(AlSi_3O_8)$. The pyroxene mineral group consists of a suite of complex Ca, Na, Mg, and Fe aluminosilicates. Olivine is also common in basalts, consisting of the magnesium end member forsterite (Mg_2SiO_4) and the iron end member fayalite (Fe_2SiO_4). These minerals release Ca, Mg, Fe, and Na cations into solution as they weather. CO_2 dissolved in water creates a weak acid known as carbonic ($H_2O + CO_2 \rightarrow H_2CO_3$). The metal cations in solution replace the hydrogen ions to form carbonate minerals such as calcite ($CaCO_3$), magnesite ($MgCO_3$), and siderite ($FeCO_3$).

When the Hellisheidi geothermal power plant in Iceland was constructed in 2006, researchers from Reykjavík Energy, the University of Iceland, CNRS in Toulouse, France, and the Earth Institute at Columbia University in New York began investigating what would happen when CO_2 was dissolved in the spent powerplant water and injected into basalt. The studies ran field tests to inject saturated, aqueous solutions of carbon dioxide into a vesicular basalt along with spent hydrothermal water from the power plant. To the surprise of everyone, instead of requiring decades to crystallize, substantial amounts of carbonate minerals precipitated in just two years [5]. The basalt held the CO_2 in place long enough for the mineralization to occur.

The organic carbon that makes up the bulk of coal is capable of adsorbing CO_2, providing a significant amount of geologic storage. Most coal contains adsorbed methane attached to organic carbon surfaces. Because of the different molecular shapes (CO_2 is rod-like and methane is more beach-ball shaped), about three times more CO_2 than methane can adsorb onto a given surface area of coal. This displaces the methane, which is now free to migrate. Where that methane goes is problematic and can sometimes produce disastrous consequences. If it ends up in somebody's basement and blows up their house (this has happened), then the injection company could face major liabilities. Even if the displaced methane harmlessly makes its way to the surface and then into the atmosphere, it is a greenhouse gas with 86 times the global warming potential of CO_2 over 20 years, and 28 times the potential of CO_2 over a century. Another problem is that the adsorption process can cause the coal to swell, cutting off flow paths and making it difficult for large amounts of CO_2 to enter the seam. Injection of carbon dioxide in coal samples has been investigated in the laboratory, and there have been some field tests to inject CO_2 into coal. Because of the complications with methane, however, it is not especially favored for carbon dioxide storage.

Shale formations have been extensively drilled and developed for gas production in the first two decades of the twenty-first Century. Like depleted oil fields, the potential for using existing wells to inject carbon dioxide is economically attractive. Organic rich shales also contain an adsorbed gas component like coal seams, allowing for a significant additional storage volume beyond just pore space. Shale has a dual-porosity structure, with the bulk of the pore volume consisting of very small void spaces between mineral grains, within mineral structures, and within organic components. The second part of the pore system is composed of natural fractures and the manufactured hydraulic fractures. The fractures make up less than 1% of the pore volume but provide high permeability flow paths for the gas to move through the rock. This dual porosity structure is why shale is noted for producing a lot of gas early on as fractures drain, followed by very long term production at much slower rates as gas migrates out of tiny pores. An assessment of shale gas resources in the United States that could be used for carbon dioxide storage came up with some impressive numbers, but injecting CO_2 into shale has the opposite problem of getting natural gas out [6]. The carbon dioxide will quickly fill the fracture system, and then slowly and painstakingly make its way into the tiny pore spaces. Working with shale is an exercise for the steady, patient, and committed. The shale gas industry has been strongly opposed to any injection of CO_2 into productive gas shales. They have refused to allow field experiments on their shale wells, likely due to concerns that CO_2 injection could foul the reservoir and reduce its future gas productivity. Until access to depleted shale wells is granted, understanding how carbon dioxide might be sequestered in shale remains limited to lab experiments and computer models.

Porous sedimentary rocks at great depths contain saltwater rather than the fresh groundwater found in shallower aquifers near the surface. Fresh water from rain and melted snow on the surface that infiltrates into the ground to recharge shallow aquifers only gets down to moderate depths, and porous rocks at depths below about 500 meters (1640 ft.) typically do not receive the freshwater recharge. These deeper rocks contain saltwater, known as connate water, which is often the residual seawater that was trapped in the sediment when it was deposited in an ocean or inland sea. Deep porous rock units can be used to sequester substantial amounts of CO_2 under high pressure. The biggest obstacle is the cost of drilling deep wells, which is not a trivial expense, especially at great depths. Tracking the potential migration of the carbon dioxide at great depths is also extremely challenging. Deep saline aquifers could potentially be used in some situations where there are no other choices for sequestration, but shallower options are often better and cheaper.

Terrestrial Sequestration Most terrestrial sequestration ideas revolve around planting trees and using soils and biomass in forests to sequester CO_2 as organic carbon. The amount of carbon dioxide actually being stored is debatable because the numbers on this are just not solid. Trees hold carbon dioxide but they grow slowly. A forest can take decades to mature. In areas that have been de-forested, re-planting with commercial timber may restore the "forest," but it is not as efficient at sequestering carbon as the soil and biomass in the old-growth forest ecosystem.

Sequestration in soils often utilizes a substance called biochar to add carbon to the soil. Biochar is "the solid black carbon material obtained from the thermochemical conversion of biomass in an oxygen-limited environment" and is nothing more or less than charcoal. If mixed into the soil instead of burning it to cook hamburgers, it will sequester the carbon for many years. Both biomass and soil sequestration have serious issues regarding their effectiveness in terms of scale-up. However, any and all methods will help to store carbon dioxide, and while planting trees may not be the single solution to carbon sequestration, it can certainly contribute.

The concept of the circular economy incorporates CO_2 into durable materials where it can be turned into useful products and chemicals. One idea is to use genetically engineered microbes to create graphene, carbon fiber, carbon nanotubes, or other desirable materials out of captured CO_2. Products could also be made using chemical engineering technology. For example, CO_2 can be used as a feedstock to make ethylene, which can then be made into a host of materials ranging from gasoline to plastic bags. None of this is quite ready for prime time at the moment because the economics of these processes remain wildly uncertain. At present, the cheapest source of ethylene is condensate produced as a byproduct of fossil natural gas production.

One option under development is to use microbes to precipitate carbonate minerals from captured carbon dioxide in a process similar that described earlier in the Iceland basalts. Precipitation of carbonate from an aqueous solution is exactly analogous to the precipitation of natural limestone. The chemical reactions would be the same, but adding microbes to the process speeds things up. The carbonate minerals could be shaped and formed into paving stones, bricks, blocks, countertops, stone facings, or virtually anything else currently made from natural limestone, marble, concrete block, or clay bricks. This "cultured limestone" would have the same properties of the natural rock, and like the natural rock, it would be composed of carbon dioxide from the atmosphere. The formation of limestone rock is nature's way of removing excess CO_2 from the air. Unfortunately, nature takes geologic time to accomplish this, so speeding up the process with microbes could be a workable solution.

Ocean Sequestration Carbon dioxide will become a supercritical fluid if injected into the ocean at depths greater than one kilometer. Supercritical CO_2 has a density greater than the surrounding seawater and will dissolve and disperse without coming to the surface. One method to sequester CO_2 in the deep oceans is to inject it from pipelines on deep water oil drilling platforms or from ships. The risks of major environmental impacts on sea creatures from injecting large quantities of CO_2 into the deep ocean are high. It could be especially dangerous to any organisms that found themselves uncomfortably close to a concentrated CO_2 injection point.

Adding CO_2 to the deep ocean water also carries the risk of a phenomenon known as "turnover." A tragic example of this occurred in 1986 at Lake Nyos in Cameroon, Africa. This lake is located in a volcanic crater where migrating gases had caused the bottom waters to become supersaturated with CO_2. On August 21 of that year, the lake's equilibrium was disturbed, resulting in a turnover of the top and bottom waters. This brought the supersaturated waters to the surface and released over a megaton of CO_2 gas. The heavier-than-air gas flowed down a valley, displaced oxygen, and suffocated 1746 villagers and thousands of livestock within 25 km (16 miles) of the lake without warning. To prevent such events, the lake now has a venting system that continuously pumps bottom waters to the surface, avoiding the buildup of dangerous levels of carbon dioxide. Climate phenomena like "El Niño," which periodically alter ocean circulation patterns and bring deep ocean waters to the surface off the Chilean coast in winter, could potentially release gigatons of sequestered carbon dioxide back into the atmosphere. Therefore, injecting CO_2 into the deep ocean without certainty of its permanence is a risky strategy.

10.2.3 Green Steel and Cement

Concrete is a versatile, durable, and popular construction material. The world-wide production of concrete is responsible for approximately 8.6% of all anthropogenic CO_2 emissions because of increased global urbanization and subsequent infrastructure construction [7].

Concrete emits substantial amounts of CO_2 during the manufacturing process and absorbs some of that CO_2 back when it cures. Concrete consists of a gravel and sand aggregate bound together with a material called Portland cement. This cement is made by heating crushed limestone, clay and several other ingredients in a kiln to a temperature of more than 1425 °C (2600 °F). The calcium carbonate ($CaCO_3$) making up the limestone breaks apart or "slakes" under this high temperature, forming CO_2 gas and solid calcium oxide (CaO) through a process called calcination. Calcium oxide is used as a component of the cement, but the CO_2 gas created during calcination is considered a waste product and is typically vented to the atmosphere. The kiln itself may be an additional source of CO_2 emissions if it is heated using fossil fuel (as most are) and the combustion products are vented to the atmosphere.

As concrete cures, atmospheric CO_2 reacts with the CaO in the Portland cement to re-generate calcite or similar carbonate minerals that make concrete as strong as a rock. Although fresh concrete will harden in hours, a full cure to completely mineralize the cement takes weeks to months. If the cure is done properly, essentially all of the CO_2 released from the cement calcination process will be re-absorbed by the concrete. In this sense it is carbon neutral, except for the emissions from heating the kiln during the production process.

Installing CCS to capture CO_2 emissions from fossil fuel combustion will make concrete carbon-neutral, and if the CO_2 emissions from the limestone slaking process are captured and sequestered as well, the concrete will become carbon negative as it absorbs CO_2 from the air while curing. This process could help reduce atmospheric greenhouse gas levels.

The cement curing reaction can be enhanced by introducing CO_2. This can be achieved by dissolving carbon dioxide into the water that is used to make the concrete flow like a liquid when poured, leave channels in the concrete that allow air to enter, or by forcing CO_2 into the hardened but not yet cured concrete under pressure. Adding CO_2 helps to improve the cure and produces a stronger, more durable material.

The global iron industry currently produces around two billion tons of steel each year, along with more than three billion tons of CO_2, mostly from blast furnaces. There are a number of improvements to the steelmaking process that can reduce emissions. The first is obviously to capture and sequester the CO_2 carrying off the oxygen from the iron ore instead of allowing it to vent to the atmosphere. A second improvement would be to use a carbon source that is not derived from fossil fuel like coke. Biofuel carbon such as charcoal or biogas are carbon neutral and do not add to the climate burden. A third improvement would be to increase the use of electric arc furnaces over blast furnaces, recycle more scrap steel, and use more directly reduced iron. All phases of the steel industry could benefit from efficiency improvements in the use of energy and materials.

One of the most significant improvements to steel production would be to forgo the use of carbon-based reductants altogether in favor of hydrogen-based reductants. Hydrogen readily combines with oxygen and forms water. Steelmakers in Sweden have been working on a venture using green hydrogen to make steel at a pilot plant. Substituting hydrogen for carbon reductants could cut the steel industry's emissions by around 90% and make a significant dent in global greenhouse gas emissions.

10.3 The Future Role of Artificial Intelligence in Renewable Energy

The global shift toward renewable energy is not only a technological necessity but a planetary imperative. As the demand for decarbonization accelerates, the renewable energy sector is undergoing a rapid transformation, one that is increasingly driven by data, complexity, and the need for adaptability. Artificial Intelligence (AI), with its ability to learn from vast datasets, identify patterns, and make autonomous decisions, is emerging as the digital backbone of this transformation.

While AI is already playing a pivotal role in areas such as weather forecasting, load prediction, and preventive maintenance, its **future applications** are set to be far more revolutionary. Rather than simply supporting renewable energy systems, AI will become central to **designing, operating, optimizing, and democratizing** the energy systems of tomorrow.

10.3.1 From Reactive Systems to Proactive Intelligence

Historically, energy systems have been reactive, responding to human inputs and basic control rules. The next generation of renewable infrastructure will be proactive. AI will enable systems that *predict demand, anticipate risks, and autonomously optimize performance* in real time.

For instance, future wind and solar farms will be equipped with AI-driven controllers that analyze microclimate data, historical performance, and market prices to dynamically adjust operations. These systems may even incorporate **self-healing capabilities**, detecting and correcting anomalies without human intervention. The goal is to create **resilient, intelligent ecosystems** that require minimal oversight and offer maximum reliability.

10.3.2 The Emergence of the Energy Internet

The renewable energy landscape is shifting from centralized power plants to a highly distributed network of generation and consumption points, solar rooftops, electric vehicles, home batteries, and microgrids. Managing this complex, decentralized ecosystem demands an equally distributed layer of intelligence.

AI will serve as the **coordinating force behind the "Energy Internet"**, a decentralized, digitized energy system that mirrors the structure of the internet. Through techniques such as multi-agent reinforcement learning, edge computing, and federated learning, AI systems will facilitate real-time communication and coordination between billions of connected energy devices.

Rather than relying on central operators, these AI agents will make *local decisions that align with global system goals*, such as reducing peak load, increasing renewable penetration, or maximizing community energy sharing. This vision will blur the traditional line between energy producers and consumers, enabling the rise of **prosumers** who both generate and use energy intelligently.

10.3.3 Climate-Resilient Energy Systems

As the effects of climate change become more pronounced, energy systems must evolve to withstand extreme weather, shifting resource availability, and rising risks. AI will be instrumental in developing **climate-resilient renewable energy systems** that adapt proactively to new environmental realities.

Future AI models will incorporate *climate forecasts, wildfire data, flooding risk maps, and sea-level projections* into energy planning and real-time operation. For example, a solar farm in a wildfire-prone area could preemptively reduce output and activate protective measures when AI predicts high fire risk. Wind farms in coastal zones might modify operations in anticipation of hurricanes or storm surges.

This shift from static design to **dynamic resilience** is essential for ensuring uninterrupted power and safety in a warming world.

10.3.4 Generative AI in Infrastructure Design

Designing efficient renewable energy infrastructure is a complex optimization problem involving thousands of variables like sun path, terrain, grid connectivity, installation cost, environmental constraints, and more. AI, particularly generative algorithms and evolutionary computation, will transform how we design these systems.

Future tools will allow developers to input high-level goals such as lowest cost per kWh or minimal ecological impact, and receive **automatically generated layouts** for solar fields, wind parks, or hybrid systems. These generative AI tools will simulate and optimize across millions of permutations in minutes, something impossible with traditional engineering methods.

Moreover, such tools will account for *future uncertainties*, designing systems that perform well under multiple possible climate, policy, or economic scenarios, rather than optimizing for just one set of assumptions.

10.3.5 AI-Augmented Energy Policy and Planning

Beyond engineering, AI will play a growing role in **guiding energy policy**. Governments and regulators will use AI-driven simulations to forecast how carbon pricing, subsidies, or grid regulations will affect emissions, affordability, and equity.

Advanced agent-based models will simulate the behavior of millions of actors like utilities, consumers, investors, under different regulatory regimes. This will allow for **evidence-based policymaking** that minimizes unintended consequences and enables adaptive governance.

In long-term energy planning, AI will help identify **least-cost decarbonization pathways**, balancing renewables, storage, transmission investments, and demand-side measures in a dynamic, data-driven framework.

10.3.6 Inclusive and Equitable Energy Access

Perhaps one of the most transformative roles of AI in the future will be its contribution to **global energy equity**. Today, over 700 million people still lack access to electricity. AI can help bridge this gap by enabling cost-effective, data-informed expansion of clean energy in underserved areas.

For instance, satellite imagery and machine learning can be used to *map energy poverty*, assess off-grid potential, and design microgrid networks tailored to local needs. AI-based credit scoring can support innovative financing mechanisms such as pay-as-you-go solar, allowing low-income households to adopt renewable energy without large upfront costs.

In this context, AI will not just enhance efficiency, but enable **just and inclusive energy transitions**.

10.3.7 Convergence with Emerging Technologies

AI's power will be amplified when integrated with other frontier technologies:

- **Blockchain** can provide secure, decentralized energy trading and data verification.

- **IoT (Internet of Things)** will supply real-time data from billions of devices, enriching AI models with hyperlocal context.
- **Quantum computing**, once mature, will solve previously intractable optimization problems in grid balancing and energy routing.

Together, these technologies will form an **intelligent, decentralized, and adaptive energy ecosystem**, capable of managing complexity, scaling globally, and meeting the world's climate goals.

In the coming decades, AI will no longer be a peripheral tool, it will be as foundational to energy infrastructure as copper wires and silicon panels. It will serve as the nervous system of a complex, distributed, and dynamic energy world, enabling renewable sources to be deployed faster, integrated smarter, and managed more equitably. However, progress in this area may face setbacks due to changing federal policies, as exemplified by challenges within the Western Energy Coordinating Council (WECC), though individual states may accelerate adoption independently.

As discussed in Chap. 8, AI offers immense potential to revolutionize the energy sector. The realities of policy and regulatory hurdles, however, highlight the complex landscape these innovations must navigate to achieve widespread adoption and impact.

10.4 Challenges and Opportunities in Emerging Markets

The global shift toward renewable energy is a defining feature of the twenty-first century, but the pace and pathway of this transition vary widely across regions. Emerging markets, defined as low- and middle-income countries undergoing rapid economic development, occupy a unique position in this energy evolution. These countries often possess some of the world's richest untapped renewable resources like abundant sunlight, expansive wind corridors, and flowing river systems, yet face distinct and multi-layered barriers that hinder renewable energy deployment. At the same time, these barriers open the door to innovative financial models, transformative policy frameworks, and inclusive development strategies that can make the energy transition in these regions both just and impactful.

10.4.1 Access to Finance and Investment

10.4.1.1 The Challenge

Financing renewable energy in emerging markets remains a central obstacle. Investors are often wary due to perceived risks, including political and regulatory instability, currency devaluation, limited grid infrastructure, and weak legal enforcement mechanisms [8]. Traditional lenders may shy away from long-term infrastructure investments in regions where creditworthiness is low and transaction costs are high.

In addition, many renewable energy technologies, particularly wind and solar, require significant upfront capital investment, which can be prohibitive for local developers without access to affordable debt or equity financing. The result is a financing gap that often leaves high-potential clean energy projects stalled at the concept stage.

10.4.1.2 The Opportunity

Despite these hurdles, emerging markets offer some of the fastest-growing investment opportunities in clean energy. Development finance institutions (DFIs), multilateral development banks (MDBs), and green climate funds are playing an increasingly important role in de-risking renewable energy investments through blended finance, concessional loans, and guarantees. For instance, the Green Climate Fund has invested billions in scalable renewable projects in Africa and Asia, leveraging private capital while improving energy access [9].

Innovative mechanisms such as green bonds, microfinance schemes, and pay-as-you-go solar services are also gaining traction. These models allow households to access distributed solar systems without high upfront costs, repaying over time based on usage, a model especially impactful in rural, off-grid regions.

10.4.2 Policy and Regulatory Support

10.4.2.1 The Challenge

A robust policy environment is essential for scaling renewable energy, yet many emerging economies struggle with policy inconsistency, lack of transparency, and poor institutional capacity. Regulatory frameworks may lack long-term targets, clear procurement mechanisms, or incentives for private investment. Unpredictable policy changes, such as retroactive tariff cuts or delays in power purchase agreements (PPAs), can further deter international developers and financiers.

Grid infrastructure is another critical bottleneck. Many developing countries have limited interconnection capabilities, unreliable transmission networks, or monopolized utilities that resist decentralized generation.

10.4.2.2 The Opportunity

When governments prioritize renewable energy through clear and consistent policy action, the impact is profound. India's National Solar Mission and auction-based tariff system has propelled it into one of the largest solar markets globally. Similarly, Brazil's renewable-friendly auction mechanisms and grid access reforms have attracted billions in wind energy investment [10].

Emerging markets can also leapfrog legacy systems by integrating distributed energy resources (DERs), smart metering, and mini-grids tailored to local needs. Regulatory reforms that enable independent power producers (IPPs), streamline permitting, allow net metering, and support feed-in tariffs can catalyze private sector engagement and technology transfer.

10.4.3 Community Engagement and Local Ownership

10.4.3.1 The Challenge

Technological and financial success alone cannot guarantee the sustainability of renewable energy projects. Many initiatives in emerging markets have faltered due to inadequate community engagement, resulting in social resistance, land disputes, or project delays. Historically, energy projects were often imposed through top-down planning, with little regard for local knowledge, consent, or benefit sharing.

Marginalized groups, especially women and Indigenous communities, are frequently excluded from decision-making processes, even when their land or resources are impacted.

10.4.3.2 The Opportunity

Inclusive and participatory renewable energy development fosters long-term success. Community-owned energy cooperatives, participatory planning frameworks, and community benefit agreements (CBAs) help ensure that local populations are not only passive recipients but active stakeholders in energy transitions [11]. Projects that provide jobs, education, training, and revenue-sharing opportunities can create socio-economic upliftment alongside environmental gains.

For example, in East Africa, solar microgrids operated with community participation have delivered not just electricity but also improvements in health, education, and local enterprise. In Latin America, Indigenous groups are co-developing wind farms, with a focus on cultural preservation and economic equity.

By integrating local governance structures and traditional ecological knowledge into project planning, developers can align energy goals with local priorities, ensuring a just and resilient energy future.

Conclusion A Dual Frontier of Challenge and Promise

Emerging markets are not merely beneficiaries of global energy innovation, they are laboratories of transformation. While financing gaps, policy inconsistencies, and socio-political complexities pose significant challenges, they also open space for pioneering solutions, inclusive models, and cross-sector collaboration.

Strategic engagement by governments, international institutions, developers, and civil society can unlock the vast renewable energy potential of these regions. In doing so, emerging markets can leap ahead, not just catching up with the global north but potentially leading it toward more equitable, distributed, and sustainable energy systems.

References

1. U.S. Energy Information Administration, Natural gas annual, September 2023 (2023). https://www.eia.gov/naturalgas/annual/. Accessed 11 May 2025
2. H. Yamada, Amine-based capture of co_2 for utilization and storage. Polymer J. **53**, 93–102 (2021)
3. L. Hong, J. Jain, V. Romanov, C. Lopano, C. Disenhof, A. Goodman, S. Hedges, D. Soeder, S. Sanguinito, R. Dilmore, Factors affecting the interaction of co_2 and ch_4 in marcellus shale from the appalachian basin. J. Unconventional Oil Gas Resour. **14**, 99–112 (2016)
4. S.Ó. Snæbjörnsdóttir, E.H. Oelkers, K. Mesfin, et al., The chemistry and saturation states of subsurface fluids during the in situ mineralisation of co_2 and h_2s at the carbfix site in sw-iceland. Int. J. Greenhouse Gas Control **58**, 87–102 (2017)
5. J.M. Matter, M. Stute, S.Ó. Snæbjörnsdottir, et al., Rapid carbon mineralization for permanent disposal of anthropogenic carbon dioxide emissions. Science **352**(6291), 1312–1314 (2016)
6. J.S. Levine, I. Fukai, D.J. Soeder, G. Bromhal, R. Dilmore, G. Guthrie, T. Rodosta, S. Sanguinito, S. Frailey, C. Gorecki, W. Peck, A. Goodman, U.S. doe netl methodology for estimating the prospective co_2 storage resource of shales at the national and regional scale. Int. J. Greenhouse Gas Control **51**(8), 81–94 (2016)
7. S.A. Miller, A. Horvath, P.J.M. Monteiro, Readily implementable techniques can cut annual co_2 emissions from the production of concrete by over 20%. Environ. Res. Lett. **11**, 074029 (2016)
8. World Bank, Financing renewable energy in emerging markets: addressing perceived risks (2023). Accessed 11 May 2025
9. IRENA, *Innovation Outlook: Offshore Wind* (International Renewable Energy Agency, Masdar City, 2022)
10. REN21, Renewables 2023 global status report (2023). Accessed 11 May 2025
11. M. Yadoo, H. Cruickshank, Community-based renewable energy: the role of cooperatives in rural electrification. Energy Policy **38**(6), 2943–2950 (2010)

Chapter 11
Case Studies of Successful Hybrid Renewable Energy Systems

11.1 Solar and Battery Integration in California

The integration of solar and battery storage technologies has become increasingly significant in high-penetration renewable energy markets, such as California. This section explores a successful project that combines solar energy with battery storage, addressing the challenges of grid reliability and energy storage while showcasing the technical solutions implemented to ensure a resilient energy supply. Through a case study in California, we examine the project's objectives, technical challenges, and performance outcomes, along with the lessons learned for future renewable energy projects.

11.1.1 Project Overview and Objectives

California has set ambitious renewable energy goals, including achieving 100% clean energy by 2045, with a heavy reliance on solar power. To address the intermittency of solar energy, especially during periods of high demand or low sunlight, the integration of energy storage systems (ESS) has become critical.

A key example of this integration is the **Copper Mountain Solar Facility**, located in California's Mojave Desert, which was designed to include large-scale battery storage systems alongside solar energy generation.

- **Project Design:** The Copper Mountain Solar and Storage project, developed by **Pattern Energy**, integrates **750 MW of solar capacity** with **300 MW of lithium-ion battery storage**. The project is designed to provide reliable, dispatchable energy to California's grid, smoothing out the variability of solar power by storing excess energy during sunny periods for use during peak demand

D. Gupta et al., *The Roadmap to Renewable Energy*,
https://doi.org/10.1007/978-3-032-09609-8_11

Fig. 11.1 Copper mountain solar and storage [3]

times, such as in the late afternoon and early evening when solar energy is no longer available [1, 2].

- **Primary Objectives:** The main objectives of this project were:
  - To enhance grid stability by balancing supply and demand through energy storage.
  - To provide energy storage that can help integrate renewable energy into the grid without causing instability.
  - To meet the state's growing demand for clean energy and reduce reliance on fossil fuels.

The project was developed to demonstrate the synergy between solar and storage, providing a blueprint for future renewable energy projects in high-penetration markets like California (Fig. 11.1).

11.1.2 Technical Challenges and Solutions

While the integration of solar and battery systems is a powerful solution for renewable energy, it comes with several technical challenges, particularly in the context of California's complex energy grid.

- **Intermittency of Solar Power:** Solar energy is highly dependent on weather conditions and daylight hours. One of the key challenges of integrating solar with battery storage is ensuring that the storage system can capture excess energy during the day when solar output is high and discharge that energy during times of low or no solar generation.

  Solution: The use of advanced forecasting models and energy management software helped to predict solar generation patterns, allowing for the optimized charging and discharging of the battery storage system. This allowed the battery to store energy when the solar output was higher than demand and release energy during periods of high demand or low solar generation [2].

- **Grid Stability and Reliability:** With an increasing proportion of the grid's energy coming from intermittent renewable sources like solar, maintaining grid stability is a major concern. The challenge is balancing variable generation with flexible demand while ensuring reliable power delivery.
  Solution: The Copper Mountain project incorporated **grid-forming inverters** and **advanced energy management systems (EMS)**. These systems enable the battery storage to work in coordination with solar generation to provide a stable, reliable power output. Additionally, the facility was designed to provide **frequency regulation**, ensuring that energy is quickly dispatched to the grid to maintain voltage and frequency levels within acceptable ranges [4].
- **Energy Storage Costs and Efficiency:** The high capital costs associated with energy storage systems, along with the challenge of maximizing their efficiency and lifespan, were key concerns.
  Solution: The project leveraged **lithium-ion battery technology**, which has proven to be cost-effective and efficient in energy storage applications. The combination of economies of scale in manufacturing and technological advancements in lithium-ion batteries helped to reduce costs while improving system efficiency. Additionally, advanced **battery management systems (BMS)** were deployed to optimize the battery life and ensure that the system could be operated at peak efficiency for longer periods.
- **Regulatory and Interconnection Issues:** Navigating the permitting and regulatory landscape in California was another challenge. Securing the necessary approvals and ensuring the system complied with California's stringent renewable energy standards required careful planning and coordination.
  Solution: The project developers worked closely with regulatory bodies, including the **California Public Utilities Commission (CPUC)**, the **California Independent System Operator (CAISO)**, and the **Federal Energy Regulatory Commission (FERC)**, to ensure compliance with local and state regulations. The project's interconnection to the grid was carefully planned to minimize disruptions and maximize grid stability.

Performance Outcomes

Results of the Integration and Insights for Future Projects The Copper Mountain Solar and Storage project successfully integrated solar generation with battery storage, delivering significant outcomes in terms of energy reliability and grid stability.

- **Operational Efficiency:** The project has proven to be a reliable source of clean energy, with the battery storage system efficiently storing excess solar energy and dispatching it when needed. The integration has helped to reduce California's reliance on fossil fuel-based peaking plants, especially during periods of high demand or low solar generation.

- **Grid Stability and Dispatchability:** By providing stored energy during high-demand periods, the project has helped to maintain grid stability in the face of increased renewable energy penetration. The ability to dispatch energy from the storage system on demand has enhanced the grid's flexibility and reliability.
- **Cost Savings and Emissions Reduction:** The project has helped reduce energy costs by providing an alternative to expensive fossil fuel-based electricity during peak hours. Additionally, it has contributed to California's greenhouse gas reduction goals by displacing fossil fuel generation with clean, renewable energy.
- **Scalability and Replication:** The success of the Copper Mountain Solar and Storage project has provided valuable insights for future projects. The scalability of the solar-storage integration model is promising, especially in other high-penetration markets like Texas, Nevada, and Arizona, where solar energy is abundant, and energy storage can provide significant benefits in terms of grid stability and reliability.

Lessons Learned

1. **Advanced Energy Management is Key:** The use of energy management systems to optimize solar generation and storage is critical for ensuring that the battery storage is charged and discharged at the right times to maximize efficiency and reduce costs.
2. **Collaboration with Regulators is Essential:** Early and continuous engagement with regulatory bodies ensures smooth permitting and compliance processes, reducing delays and ensuring the system's long-term operational success.
3. **Battery Storage Technology is Evolving:** As battery technology continues to improve, future projects can benefit from lower costs, longer lifespan, and better efficiency, making the integration of solar and storage even more cost-effective.
4. **Scalability is Key to Future Success:** The project's success demonstrates the potential for replicating the model in other regions and expanding capacity to meet increasing demand for clean, reliable energy.

11.2 Hybrid Renewable Projects in Emerging Markets

Hybrid renewable projects combining solar, wind, and battery storage have become a powerful solution to the challenges of renewable energy intermittency and grid stability, especially in emerging markets. Countries in Latin America, Asia, and Africa, where energy access and infrastructure development are key issues, are increasingly turning to these hybrid systems. By blending multiple renewable sources, hybrid projects can provide a reliable and sustainable energy supply. This section explores the hybrid renewable projects in **Latin America** and **India**, focusing on their technical integration, socio-economic impact, and long-term sustainability.

11.2.1 Project Overview and Regional Context

Hybrid renewable projects offer a compelling solution to energy needs in regions rich in renewable resources but facing challenges in energy distribution and grid stability. These projects typically combine solar, wind, and battery storage systems, capitalizing on the complementary nature of these resources. For example, solar energy is often abundant during the day, while wind generation can be strongest in the evenings or overnight, while batteries help manage fluctuations in power generation.

- **Latin America:** Countries like **Chile**, **Mexico**, and **Brazil** have made significant strides in hybrid renewable energy projects. These regions have abundant solar and wind resources, and many are moving towards large-scale hybrid projects to meet their energy demands sustainably. The **Cerro Dominador** hybrid project in Chile (mentioned earlier) is an excellent example, combining solar, wind, and battery storage to stabilize the grid and reduce reliance on fossil fuels [5].
- **India:** India is one of the largest emerging markets for renewable energy, with aggressive goals for achieving 500 GW of renewable energy capacity by 2030. The country's renewable energy mix, which includes wind, solar, and increasingly battery storage, is rapidly evolving. Hybrid projects are seen as a viable way to ensure energy reliability and meet the growing demand for clean energy. India's **Palladam Wind-Solar Hybrid Project** is a notable example. This project, located in Tamil Nadu, integrates 50 MW of wind energy and 50 MW of solar energy, backed by a battery storage system to ensure continuous power supply [6].
- **Asia-Pacific:** Other emerging economies in the **Asia-Pacific region**, such as **Vietnam**, **Thailand**, and **Indonesia**, are exploring hybrid projects. The combined strength of solar and wind is particularly beneficial in these regions, where energy demand is growing rapidly, and grid infrastructure remains underdeveloped. Vietnam's focus on solar in the south and wind in the north, for instance, has led to hybrid project exploration as a way to stabilize supply across seasons.

11.2.2 Technical Integration of Solar, Wind, and Battery

How Solar, Wind, and Battery Systems Were Integrated Hybrid renewable projects integrate various energy sources like solar, wind, and battery storage, in ways that maximize energy generation efficiency and provide grid stability. The technical aspects of these integrations are critical to ensuring reliable performance, particularly in regions with variable weather patterns.

- **Solar and Wind Synergy:** The complementary nature of solar and wind resources is a key factor in hybrid system success. In both **Chile** and **India**, solar generation typically peaks during the day, while wind power is often strongest at night or during off-peak hours. By integrating these two sources, hybrid systems ensure a more consistent energy supply throughout the day and night.

- **Battery Storage for Flexibility:** Batteries play a crucial role in hybrid renewable systems by storing excess energy generated during times of high production and releasing it during periods of low generation. The **Cerro Dominador** project in Chile, for example, uses a **150 MW/200 MWh battery system** to store solar and wind energy, helping to provide reliable energy when production dips (Cerro Dominador [5]). In India, the **Palladam Wind-Solar Hybrid Project** incorporates a battery system that stores excess energy produced during peak generation periods, ensuring continuous supply to the grid.
- **Energy Management Systems (EMS):** Advanced EMS platforms are essential for managing hybrid systems. These platforms track weather patterns, grid demands, and the operational status of each energy source. In India, for example, the **Indian Renewable Energy Development Agency (IREDA)** has been implementing EMS solutions in hybrid systems to optimize performance and reduce operational costs [7].
- **Grid Integration:** The integration of hybrid systems with the existing grid infrastructure is a complex process, especially in regions where grids are less advanced. In **Chile**, the **Cerro Dominador** [5] project is integrated into the **Central Interconnected System (SIC)**, helping to ensure reliable distribution to both urban and remote areas (Cerro Dominador [5]). In India, hybrid projects like **Palladam** contribute to improving grid stability in regions with high energy demand and help balance the intermittent nature of renewable power.

11.2.3 Socio-Economic Impact and Sustainability

The Impact of Hybrid Projects on Local Communities and Long-Term Sustainability Hybrid renewable energy projects bring substantial socio-economic benefits, including job creation, improved energy access, and environmental sustainability. These projects are often implemented in rural or remote areas, where they can stimulate local economies and contribute to sustainable development.

- **Job Creation and Local Economic Growth:** Large-scale hybrid renewable projects create thousands of jobs during the construction and operational phases. The **Cerro Dominador** project employed over 1000 workers during its construction, with many coming from nearby communities. Similarly, India's **Palladam Wind-Solar Hybrid Project** has created numerous local jobs, boosting the local economy [7].
- **Energy Access and Social Equity:** Hybrid projects provide reliable, affordable electricity to underserved communities, especially in rural areas. In both **Chile** and **India**, hybrid systems are improving electricity access, reducing energy poverty, and contributing to more equitable development. In **India**, hybrid

systems have been particularly beneficial in remote regions with limited access to the national grid.

- **Environmental Sustainability:** By combining renewable sources like wind and solar with battery storage, hybrid systems reduce the reliance on fossil fuel power plants, significantly lowering carbon emissions. The **Cerro Dominador** project, for instance, is expected to prevent the emission of around **1.5 million tons of CO_2 per year** [5]. In India, projects like **Palladam** help to reduce emissions, contributing to the country's renewable energy targets and supporting its efforts under the **Paris Agreement**.
- **Long-Term Resilience:** Hybrid renewable energy projects are not only sustainable but also resilient. They are scalable, allowing for future expansion as energy demand increases. This ensures that such projects can continue to contribute to energy security and economic growth for many years. In **Chile**, hybrid systems are seen as critical to meeting the country's ambitious renewable energy goals, and in **India**, the integration of battery storage systems is seen as a way to ensure grid stability in the long term.

11.3 Geothermal and Solar Hybrid Systems

Hybrid solar-geothermal systems, sometimes referred to as solar assisted geothermal energy or SAGE, are designed to boost the temperature of a marginal geothermal resource with solar energy, usually collected in focused parabolic troughs or dishes. These systems are generally configured in one of two ways: either the solar heat is added to the hot water or steam on the way to the power plant, or the solar heat is added to groundwater recirculated into the aquifer and the rock mass underground is used to accumulate and store the heat.

11.3.1 Overview of Hybrid Geothermal-Solar Projects

The idea for adding a solar heating loop to geothermal power plants has been around for several years, but implementing this has been challenging. The concept is to boost the temperature of the produced water with concentrated solar power (CSP), usually supplied by a parabolic reflector that focuses sunlight on a pipe or conduit containing the working geothermal fluid. The first SAGE system proposed for a geothermal power plant was designed in 2006 for the Cerro Prieto Geothermal Power Plant in Baja California near Mexicali, Mexico, but to date it has not been built. These hybrid setups work best on flash steam plants for conventional hydrothermal resources, and on lower to medium temperature geothermal resources where the working fluid is in the liquid phase and thus more efficient at absorbing heat.

The solar heat can be transferred to the fluid on the way to the geothermal power plant, in a secondary heating loop of the residual liquids from a flash steam plant, or a paired injection-production well design can be used to circulate the solar-heated fluid through an underground aquifer that will accumulate and store the heat in the subsurface rock mass for future use. The temperatures imparted to the working fluid from a parabolic solar reflector array can reportedly get quite hot—in excess of 350 °C. A solar hybrid heat loop can boost the temperature of marginal geothermal resources in the 150 °C range to more efficient levels, and it can also be used to restore heat to conventional hydrothermal systems where subsurface temperatures have declined due to production.

11.3.2 Synergy Between Geothermal and Solar Energy

Solar and geothermal energy sources complement each other technically in several ways. Solar can provide a substantial heat boost to geothermal fluids, making geothermal energy more usable over larger geographic areas. Geothermal, on the other hand, can supply steady, continuous power that is not subjected to the day-night or cloudy-day fluctuations of solar. A geologic-based geothermal system, including EGS and AGS, provides a large thermal mass to store heat, overcoming one of the major concerns about solar-thermal power systems. Rocks have a large amount of thermal inertia, and are capable of accumulating enough heat on sunny days and through the summer season that when scaled properly it can supply a geothermal power plant through the winter.

One of the most economic synergies between geothermal power and solar is the addition of concentrated solar power at an existing hydrothermal power plant to restore the heat that has been produced from the declining subsurface geothermal resource. Economic assessments show that the capital expense of adding CSP has a relatively quick return on investment by boosting the power output of the plant back to nominal design levels, and is a lower capital expense than drilling new production wells or adding extra generating units to obtain the same power capacity.

11.3.3 Case Study: Successful Implementation

A review study of geothermal-solar hybrid projects found a number of interesting design studies, but no active SAGE powerplants as of 2020. This may have changed in the intervening years, but the only announced solar hybrid geothermal power plant was by Enel Green Power North America in 2020, which intended to utilize 80,000 photovoltaic (PV) panels in conjunction with traditional hydrothermal technology to produce 24 megawatts of combined solar PV and geothermal electricity in Churchill County, Nevada. Although nominally a hybrid system, it does not meet the definition of solar assisted geothermal energy. As of this writing, no operating geothermal power plants, hybrid or otherwise, are listed on the Enel Green Power North America website for the United States.

The successful implementation of solar assisted geothermal energy could provide a tremendous opportunity for the U.S. electric power sector. The technology can be used at existing coal-fired power plants to convert them to clean, sustainable geothermal energy while utilizing existing generating, transmission, and distribution infrastructure. The boilers in a power plant don't care where the heat comes from, as long as it is hot enough. Raising the temperature of local aquifers with CSP and using that heat to generate electricity is a cost-effective, low capital expense, and quickly-implemented solution to decarbonizing electric power, possibly achieving that goal within a decade. However, like most adoptions of new technology, somebody has to be brave enough to go first.

References

1. Saur Energy International, Top 5 u.s. solar parks (2023). Accessed 11 May 2025
2. Pattern Energy, Copper mountain solar and storage project overview (2020). Accessed 11 May 2025
3. Cupertino Electric, Copper mountain solar 3 (2025). Accessed 11 May 2025
4. S. Pincetl, M. Luderer, Challenges and opportunities for solar storage integration in California. Renew. Energy **135**, 1074–1084 (2019)
5. Cerro Dominador, Cerro dominador solar and wind hybrid project overview (2020). https://www.cerrodominador.cl. Accessed 11 May 2025
6. Ministry of New and Renewable Energy, Government of India, Palladam wind-solar hybrid project: advancing clean energy in Tamil Nadu (2020). https://www.mnre.gov.in. Accessed 11 May 2025
7. Indian Renewable Energy Development Agency (IREDA), Hybrid renewable energy projects in india: a path to energy security (2021). https://www.ireda.in. Accessed 11 May 2025

Chapter 12
Conclusion and Final Thoughts

12.1 Key Takeaways from the Roadmap to Renewable Energy

In the world of renewable energy, success is no longer defined by isolated expertise but by the seamless integration of **science**, **business**, and **AI**. This book has demonstrated how these three pillars must come together for projects to be viable and sustainable in a rapidly changing energy landscape.

- **Integration of Science, Business, and AI**
  Renewable energy projects require not just scientific innovation but also strategic business acumen and data-driven decision-making. Science helps us understand energy sources, optimize technologies, and assess environmental impacts. Business is crucial in aligning market dynamics, financial viability, and stakeholder engagement. AI empowers us to evaluate vast datasets, anticipate project risks, and maximize efficiencies throughout the project lifecycle. When combined, these elements create a robust foundation for transforming ideas into successful, large-scale energy projects. As the energy industry evolves, this fusion will become increasingly essential for scaling up renewable solutions.
- **Lessons Learned from Case Studies**
  Through the real-world case studies presented in this book, we have explored a wide array of challenges and triumphs in renewable energy development. Key lessons have emerged, such as the importance of early-stage feasibility analysis, the need for adaptive management in uncertain regulatory environments, and the role of collaboration across multidisciplinary teams. These case studies remind us that while technical innovation is important, it is the ability to navigate complexities in policy, economics, and local engagement that often determines success or failure.

D. Gupta et al., *The Roadmap to Renewable Energy*,
https://doi.org/10.1007/978-3-032-09609-8_12

- **Holistic Approach to Project Development**
  One of the central themes in this book has been the importance of a holistic approach to renewable energy project development. No single discipline or strategy can independently deliver the desired outcomes. Rather, a comprehensive approach, one that draws upon the expertise of scientists, engineers, project managers, financial analysts, and stakeholders, is essential. A multidisciplinary team working together ensures that all aspects of a project, from design and financing to social impact and sustainability, are carefully considered. This holistic perspective encourages innovation, mitigates risks, and accelerates progress toward a sustainable energy future.

12.1.1 The Future of Interdisciplinary Approaches

The future of renewable energy lies in the continuous evolution of interdisciplinary approaches. As the world faces the urgency of transitioning to a sustainable energy model, we are entering an era where technologies, markets, and policies must come together in ways we have not yet seen. For projects to succeed on a global scale, it will be essential that the fusion of science, business, and AI adapts, evolves, and accelerates. Based on my experience navigating this space, I believe this future will be shaped by three key drivers: emerging technologies, cross-sector collaboration, and innovative policy frameworks.

- **Evolving Technologies and New Opportunities**
  Renewable energy is at the cusp of a technological revolution. Every breakthrough, whether in energy storage, artificial intelligence, or materials science, represents a new opportunity for interdisciplinary collaboration. For example, I've witnessed firsthand how **AI and machine learning** are transforming grid management and predictive maintenance for renewable energy assets. These technologies allow us to analyze real-time data in ways we never could before, optimizing operations and minimizing downtime. Meanwhile, advancements in energy storage, such as **solid-state batteries**, will unlock the potential to store solar and wind energy at a much larger scale. The future will see a closer marriage between **business models** that can integrate these technologies and **scientific innovations** that bring them to life. The key opportunity here is for all sectors, science, engineering, business, and AI, to collaborate early on in product development to reduce costs, maximize efficiencies, and scale these technologies rapidly across industries.
- **Cross-Sector Collaboration**
  Having worked across various sectors in renewable energy, I've come to realize that no one player can solve the challenges we face. Cross-sector collaboration spanning industries, governments, and academia will be a defining feature of the future of energy. Governments must provide the regulatory frameworks and long-term stability that give businesses the confidence to invest. Meanwhile, the business sector must lead the charge in deploying these technologies at scale.

This will require close collaboration with **academia** for research, development, and talent cultivation, and with **governments** for ensuring that the infrastructure and policy landscapes are conducive to rapid deployment. In my experience, **early engagement with local communities and stakeholders** is just as critical as working with engineers or business leaders, ensuring that projects are not only scientifically sound but also socially accepted and impactful. We must continue to engage in dialogue across sectors to create a world where renewable energy is not just a goal but a reality that enhances economies, communities, and ecosystems.

- **Innovation in Policy and Market Mechanisms**
  When it comes to policy and market mechanisms, we are only beginning to scratch the surface of what's possible. Through my work in renewable energy development, I've seen how market incentives like **renewable energy certificates** (RECs) and **power purchase agreements** (PPAs) can drive large-scale projects. But for renewable energy to truly compete with fossil fuels and become mainstream, we need more **innovative policies** that will further reduce the financial barriers to entry. **Carbon pricing**, for instance, is already proving to be a powerful market mechanism that incentivizes companies to reduce their emissions and embrace cleaner technologies. As we move forward, it will be vital for policymakers to craft **dynamic regulatory environments** that not only support technological growth but also foster collaboration between business, science, and the wider society. By introducing tools like **green bonds**, **impact investing**, and targeted tax incentives, we can create a robust financial framework that supports long-term sustainability. These policies will encourage the next wave of technological advancements while addressing climate goals and market needs simultaneously.

In conclusion, the interdisciplinary approach that has been the cornerstone of this book will continue to evolve and expand in the coming years. The integration of science, business, and AI is no longer a luxury. It is a necessity for the development of a sustainable, low-carbon energy future. We must embrace emerging technologies, strengthen cross-sector collaborations, and innovate in policy and market mechanisms to drive the renewable energy revolution. Together, these efforts will shape the next chapter of energy development, one that is not only efficient and profitable but also equitable and enduring.

12.1.2 Call to Action for Industry and Policymakers

As we stand at the crossroads of a transformative energy future, the time to act is now. Renewable energy is not just a sector of the economy, it is the foundation for a sustainable, resilient, and equitable world. The urgency to transition from fossil fuels to clean energy sources is undeniable. However, this transition requires bold actions from both **industry leaders** and **policymakers**. To catalyze this change, we must prioritize long-term sustainability, invest in innovation and education, and foster global cooperation. These actions will lay the groundwork for a green energy future.

- **Prioritizing Sustainability in Energy Policies**
  Governments must take a leadership role in shaping energy policies that prioritize long-term sustainability over short-term gains. We have seen how effective **policy frameworks** can drive large-scale investments in clean energy technologies, create jobs, and stimulate economic growth. The next step is to develop policies that go beyond immediate market incentives and focus on the **long-term impact** of renewable energy on the environment, communities, and the economy. This means establishing **clear renewable energy targets**, incentivizing **carbon-neutral practices**, and implementing **resilient infrastructure** strategies. Policymakers must also recognize the importance of integrating social equity into these policies, ensuring that all communities, especially marginalized ones, benefit from the clean energy transition. By creating policy ecosystems that encourage continuous innovation, we can drive sustained growth in the renewable energy sector and combat climate change effectively.
- **Investing in Innovation and Education**
  Innovation in technology and **education** will be the cornerstone of a sustainable energy future. Industry leaders and policymakers alike must **invest in research and development** of next-generation technologies, from energy storage solutions to smart grid systems, that can overcome current limitations. Equally important is the investment in **education** and **training** for the workforce of tomorrow. The renewable energy sector is evolving quickly, and to stay ahead of the curve, we must cultivate a skilled workforce that can adapt to new technologies and methodologies. **Academic institutions**, **corporations**, and **governments** must partner to develop training programs that bridge the knowledge gap in areas like AI, data analytics, sustainable engineering, and energy management. Only through this investment in innovation and education can we ensure the **global competitiveness** of our energy systems and equip the next generation of leaders with the tools they need to accelerate the transition to a green economy.
- **Global Cooperation for a Green Energy Future**
  The challenge of combating climate change and ensuring a sustainable future is a **global endeavor**. No country, no sector, and no industry can do it alone. **Global cooperation** is essential in tackling climate challenges through renewable energy solutions. This means countries must come together to share knowledge, resources, and best practices, particularly in **technology transfer** and **policy development**. As a global community, we must set aside differences and focus on what unites us, the shared goal of building a sustainable future for all. This requires **international agreements** on **climate action** and a shared commitment to meet and exceed **climate targets**. **Collaboration between nations**, alongside the private sector and academic institutions, will accelerate the scaling of clean technologies and unlock new opportunities in emerging markets. Only through collective action can we create a truly **green energy future** that benefits people, the planet, and future generations.

MIX
Papier aus verantwortungsvollen Quellen
Paper from responsible sources
FSC® C105338

If you have any concerns about our products,
you can contact us on
ProductSafety@springernature.com

In case Publisher is established outside the EU,
the EU authorized representative is:
Springer Nature Customer Service Center GmbH
Europaplatz 3, 69115 Heidelberg, Germany

Printed by Libri Plureos GmbH
in Hamburg, Germany